Embedded Systems Design
The Engineer's Toolbox

Black and White Version

Embedded Systems Design
The Engineer's Toolbox
Mark Donners

Layout and artworks: Mark Donners
Editor: Evangeline Donners
Published by: Mark Donners
Year of publication: 2016

Copyright © 2016 by Mark Donners

All rights reserved. This book or any portion thereof may not be reproduced or used in any manner whatsoever without the express written permission of the publisher except for the use of brief quotations in a book review or scholarly journal.

First print: December 2016

ISBN 978-1-326-88641-7

Published by Mark Donners
Mark.Donners@judoles.nl
Hoensbroek, The Netherlands

www.judoles.nl

Ordering Information:
This book can be ordered at www.lulu.com. Also, you might want to check out your local or online bookstore.

For the terminology 'Mind Map' used in this book, take note of the following:
Mind Map (Mindmap) is a registered trademark of the Buzan Organisation Limited 1990, www.thinkbuzan.com

Dedication

To my lovely wife, Evangeline and beautiful daughter, Juliana Marie.

Thank you! Without your support and patience, I would never have achieved my dream to become a licensed engineer and a book writer.

Content

1 Foreword 11
2 Designing 15
   2.1. The need to validate 16
   2.2. Common Methods of Designing 18
   2.3. A 4-step method 23
      2.3.1. The Specify Stage 25
      2.3.2. The Design stage 31
      2.3.3. The Create Stage 36
      2.3.4. The Validate stage 37
3 The toolbox 41
   3.1. Specifications & requirements Card 'SPEC-CARD' 43
   3.2. The interview 44
      3.2.1. Preparations 45
      3.2.2. The interview 46
      3.2.3. Documenting specifications and requirements 47
   3.3. Mind mapping 48
   3.4. Use Case diagram 49
   3.5. Use Case Description 50
   3.6. Bubble diagram 51
   3.7. Context diagram 53
   3.8. Protocol diagram 54
   3.9. Activity diagram 55
   3.10. Flow diagram 56
   3.11. Nassie Schneidermann diagram 58
   3.12. State diagram 60
      3.12.1. Finite state machine 61
      3.12.2. State chart 63
   3.13. Brainstorming & Brain writing (wiki) 65
   3.14. Creative tools 68
   3.15. Morphologic table 70
4 Case study 1: Sports speed guard 71
   4.1. Specify Stage 72
      4.1.1. Preparations 73

- 4.1.2. The interview 74
- 4.1.3. Info conversion 76
- 4.1.4. Product sheet 77
- 4.1.5. Context diagram 79
- **4.2. Design Stage** **80**
  - 4.2.1. Defining the system 80
  - 4.2.2. Splitting up into partial designs 82
  - 4.2.3. Defining the partial designs 84
  - 4.2.4. Concluding the Design stage 91
- **4.3. Create Stage** **92**
  - 4.3.1. Context diagram 93
  - 4.3.2. Creating the partial designs 94
  - 4.3.3. Hardware 100
  - 4.3.4. Firmware 102
- **4.4. Validate Stage** **112**
- **4.5. Finishing the case** **116**

5 Case Study 2: Old Riddles die hard 117
- **5.1. Specify stage** **118**
  - 5.1.1. Preparations 118
  - 5.1.2. The interview 119
  - 5.1.3. Brainstorm 120
  - 5.1.4. Product sheet 121
  - 5.1.5. Converting to specifications 122
  - 5.1.6. Context diagram 123
- **5.2. Design Stage** **124**
  - 5.2.1. Defining the system 124
  - 5.2.2. Splitting up into partial designs 126
  - 5.2.3. Defining the partial designs. 127
  - 5.2.4. Concluding the Design stage 133
- **5.3. Create Stage** **134**
  - 5.3.1. Context diagram 134
  - 5.3.2. Creating the partial designs 135
  - 5.3.3. Hardware 138
  - 5.3.4. Firmware 139
- **5.4. Validate stage** **152**

6 Case 3: Embedded Sports Trainer 153
- **6.1. Specify Stage** **154**

	6.1.1.	Preparations	154
	6.1.2.	Functionality of the system	155
	6.1.3.	Product sheet	156
	6.1.4.	Specifications and requirements	157
	6.1.5.	Context Diagram	158
6.2.		**Design Stage**	**159**
	6.2.1.	Defining the system	159
	6.2.2.	Splitting up into partial designs	160
	6.2.3.	Defining the partial designs	162
	6.2.4.	Concluding the Design Stage	168
6.3.		**Next Stages, Create and Validate**	**169**
7		*The End*	*171*

1 Foreword

'There is nothing more difficult to take in hand, more perilous to conduct, or more uncertain in its success, than to take the lead in the introduction of a new order of things.'
Niccolo Machiavelli

Being an Engineer, I come in contact with all sorts of designs regularly. In the role of design engineer, I often design systems. Most designs are not too extensive and include hardware and software and sometimes mechanics. These are the so called embedded systems and mechatronics.

Living in a World in which Electronic development progresses with the speed of light, one would expect that over time, enough methods for developing embedded systems have been developed and documented. In my humble opinion, nothing is further from the truth. Most documents I was able to find over several decades were either to abstract or too specific. I only know of a few books that cover more than one topic. Up until now, including my work as an electronic engineer at one of the biggest universities of the Netherlands, I was not able to find one book that gives me enough guidelines and tools on how to design embedded system. Some books are very good in explaining the design of electronics; others do the same for software but none I found that cover everything. Although I am well aware that "everything" is a lot indeed, I do not think it is realistic to expect to find a book that can cover all.

Nevertheless, I do not want to talk about the actual design of the schematic, or the programming of a FPGA. I do want to talk about some methods that can be used to streamline the process from idea to prototype.

One of the problems I encountered in designing embedded systems is the combination of digital and analog Electronics. Even mechanics like an actuator can be involved. In those cases where the combination of hardware and software cannot be designed in one breath, the design should be split up into several smaller designs that somehow work together. This also could mean that the analog part of the design should be separated from the digital design.

To be fair, there are some methods of designing available that partly or completely include analog and digital components but they are not engineered to help you with your embedded system design.

By combining little pieces of all sorts of design methods with my own experience, I came to the design method described in this book. What I have written in this book is not the Holy Grail; I simply hope to give some guidance and useful tools to those designers out there who are starting their electronic carrier.

This book contains:

Known design methods
4 step design method
Designer's toolbox
Use Case s with examples

For the purpose of easy reading, persons in the book are introduced as 'he'. Whenever a person in this book is mentioned as being a 'he', this person could of course also be a she.

2
Designing

'Since new developments are the product of a creative mind, we must therefore stimulate and encourage that type of mind in every way possible'
George Washington Carver

2.1. The need to validate

Before getting into details about the different design methods, one should remember the bigger picture. What exactly is the process of designing? Irrespective of the design method, the purpose is always the same: to transform an idea into a product or solution well guided by the demands, wishes and specifications.

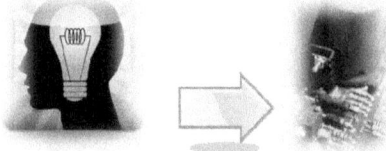

To transform an idea into a product, one has to work out the design of the actual product. To be able to design a product, one should first establish what it is to achieve and under what circumstances. This can be done by recording and documenting the demands and wishes that the customer has for this product. As a designer, it is in everyone's interest to take part in making this inventory of specifications. The designer should take part interactively because a customer does not always put in words what he really wants. He might ask for an atomic accelerator but what he really wants is a centrifuge to dry his laundry. This might sound silly but do not forget; a customer often has no specific knowledge about how things can be done. He simply has a brilliant idea.

A good designer always thinks side by side with the customer. Together, both the designer and the customer try to find a working solution. Only after the specifications are clearly documented, the designer can start with the actual process of designing. It is possible to classify the different sorts of specification. For instance: product specifications and user specifications. However, it is more important to establish which specifications are critical and which are optional. In the end, the product should meet all the critical specifications and some (or all) of the optional specifications.

Sometime it is necessary to meet each other half way in order to complete the design. This could mean that some specifications are not met. A design is only successful if both the designer and the customer are pleased in the end.

To check if the critical and optional specifications are met, the product has to be validated before it is officially handed over to the customer. Unfortunately, the validation of the specifications as agreed upon earlier in the process of designing is often forgotten. Within the enthusiasm of designing, it is so easy to get lost in a direction without noticing the side track you could be on.

It is therefore most important to keep track of your specifications throughout every step of the process and evaluating the final design against these specifications is always necessary.

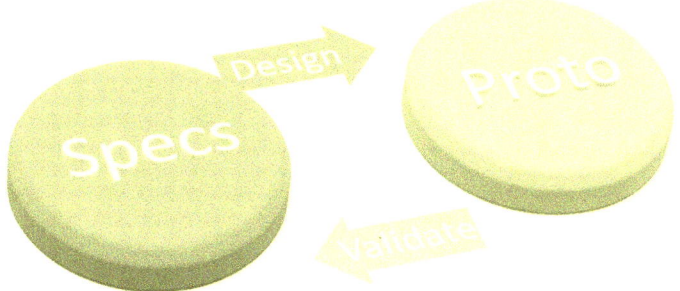

Figure 2.1 Design and validation go hand in hand.

If during the design, the designer finds out that the design is off track with the specifications, it is necessary to (partly) redesign. It is never a good option to continue the path you are on if you know you have a problem. Sometimes it just seems easier to believe that everything will work out in the end! This is of course a false assumption.

2.2. Common Methods of Designing

There are in fact many methods to develop an idea into a working solution. Most methods are based on functional decomposition. With functional decomposition the complete design is divided into several smaller sub designs that are mostly grouped by function. An example of this method can be found in the design of a doorbell intercom system. The overall design of a doorbell intercom can be divided into systems like the door relays, user interface and communication interface, to name just a few. Maybe this system includes a camera and a registration system. Dividing the overall design into smaller pieces is part of the design process and takes place before a decision has been made about what technology to use. At this point, it is unknown if we are dealing with analog electronics, digital electronics, mechatronics or any combination of those. Also, no decision has been made about firmware or software. The sub designs that have been created at this point should be considered as an individual process with their own processor. This processor is not an actual processor like a microchip but should be seen as a virtual one. It could be a microprocessor but it might as well be a human or a motor. Unfortunately this approach cannot be applied to all parts of a design. How would one, for example, describe an analog signal converter? A process with some inputs and some outputs? Yes, true! However, what would we consider being the processor in this case?
Nevertheless if this signal converter was built with digital electronics, the use of a microprocessor would be a real possibility. But what about the firmware of this processor? Could this be considered the brain of the design? It all depends!
For a microprocessor, one would have the tendency to say yes, the firmware could be considered the brain of the design. Would you also say the same if the microprocessor was replaced by a PAL or CPLD, or would this depend on the programming of this CPLD?

These unanswered questions are exactly what I am missing in common methods of designing.

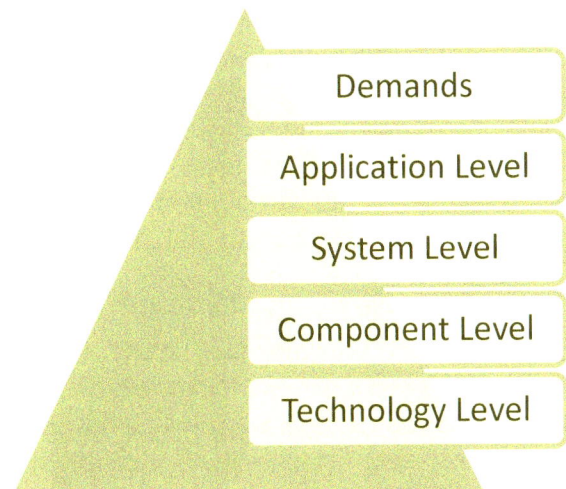

Figure 2.2 Top-down design model

A commonly used approach for several methods of designing is known for being a "Top-Down" method.
First, one takes a look at the complete design as a whole. It is similar to the way an eagle looks at earth during flight. The eagle starts at high altitude, seeing the bigger picture. However, every time the eagle circles around at a lower altitude the details of his view will increase. This is similar with the "Top-Down" method of design. The more frequent the design is looked over, the more details are engineered. Usually, the first levels from the top down are more general and the choice of what components to use has not been made at this point. The purpose of the top levels is to determine what is included in the design and what is not. It is a way to put things into context. The same levels are also used to decide what functionality will be described in more detail in the level below.

Within the top levels it is not determined what technologies or components will be used in the actual design. In fact, only in the lowest level, a decision is made about the technologies to use. This could be a CPLD, FPGA or analog electronics, to name just a few.

Because flowcharts and state diagrams are used throughout the different levels of the design, the result in the lowest level is often one of a digital nature. This is especially true if the state diagrams are already worked out in detail in the application level. This method will work fine if there is no limitation to which technology has to be used on the lowest level.

However, while designing embedded systems the designer is often influenced by experience with certain embedded systems and/or microcontrollers. Therefore, the choice of what technology to use is made sooner than on the last level. It is also possible that during the intake interview with the customer, the choice of technology is already made or the choice is limited.

For example, if the behavior of a system is worked out in detail using state diagrams in one of the top levels, it is fairly easy to generate working firmware. However, not all compilers can convert state diagrams into code that is ready to be downloaded into a microchip. In that case, the designer might convert the state diagrams into working code manually. Although this is a realistic scenario, it is also time consuming.

So, does this mean that the common "Top-Down" method is useless? No! Every designer works according to some sort of method but not all are aware of this fact. Even the designers, who claim to work without any method, will use their previous experience in the next assignment. This could also be considered a method. Let us not forget that a method of designing is only a tool to assist the designer and it is most certainly no Holy Bible.

Before I introduce my preferred method of designing, let me tell you a bit on how another popular design method works. This method consists of three levels. The third level has different stages. The first two stages are used to determine the behavior of the system by documenting the specifications. On the third layer, the actual design is divided into different levels. Each level will take the designer closer to the actual solution.

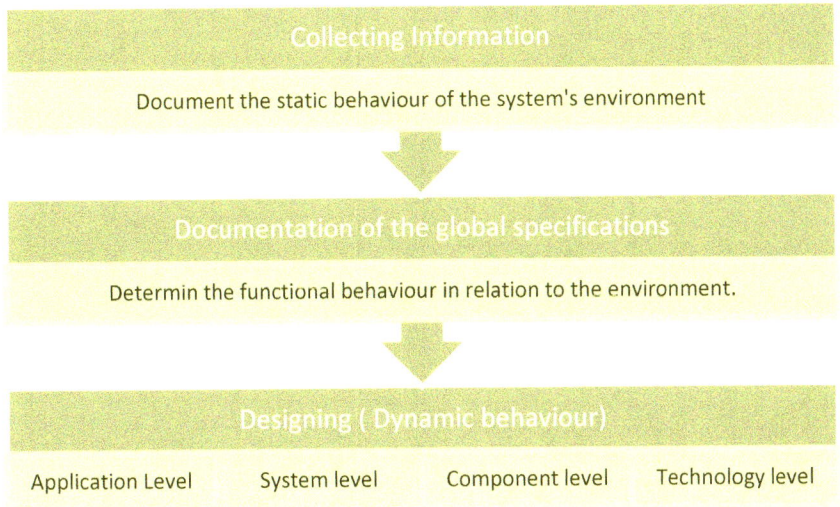

Figure 2.3 Common levels of designing

For each level, there are one or more tools available that can be used by the designer to complete the level. It is not uncommon to use certain tools during multiple levels. In that case it makes sense to increase the details on a lower level.

	Plain text	Interview	Bubble diagram (DFD)	Context diagram	Protocol diagram	Activity diagram	Flow diagram	State diagram	Truth Table	ASM-chart	Karnaugh diagram	CPLD/FPGA Design	C code	PCB Design & Schematics
Collecting Information	•	•												
Documentation of the global specifications	•		•											
Designing														
Application lvl	•			•	•	•	•							
System lvl	•		•	•				•	•	•				
Component lvl								•	•	•	•	•		
Technology lvl												•	•	•

Figure 2.4 Available tools to be used throughout the design stages

In my opinion, this "Top-down" method as I have just described, is a step in the right direction. Although, I do believe that it is less qualified for designing embedded systems, it does include some tools that are very useful. I will not elaborate about this design method because I will describe my own method next. Some methods will be included in the toolbox you can use later.

2.3. A 4-step method

The method I came up with divides the design process into 4 levels, also known as stages. The method includes iterative progress. This means that the circle of stages, as illustrated below, can be used over and over again until the design is finished. A design is finished when the validation stage confirms that all the specifications as described in the SPECIFY stage are met. If this is not the case, the circle has to be completed once more and this has to be done as often as required.

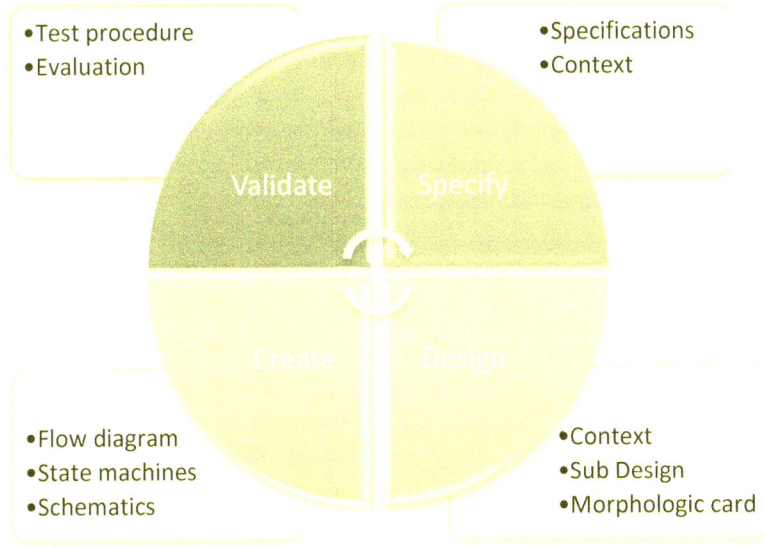

Figure 2.5 The 4-step design method

Now, let me elaborate on all stages of this design method to give a complete impression.

The diagram below shows a summary of what has to be done during each stage.

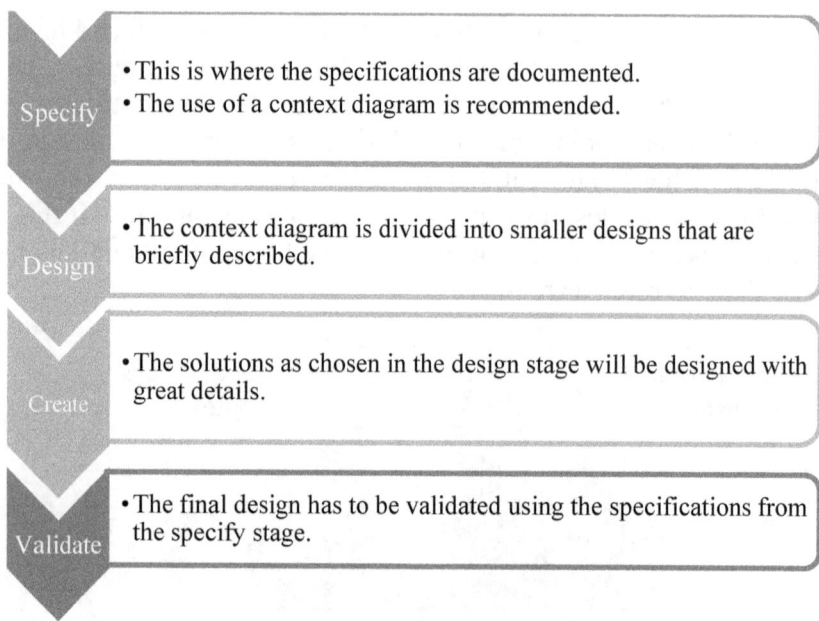

Figure 2.6 Activities during the 4 design stages

Although all stages are critical in regards to what decisions to make, the last stage will show you if your work has been successful or not. If ever, the last stage will show you that the product you just designed, does not comply with the required specifications, it is crucial to have another go at the 4 step circle. So, in that case you have to start again with your specifications and (partly) re-design using the stages, to make the design compatible with the specifications. This can be done as often as it is necessary.

2.3.1. The Specify Stage

During the SPECIFY stage, all specifications and requirements are documented. Usually, this stage starts with a conversation between designer and customer. The customer's idea or wishes are translated into a 'black box' that has its own properties.

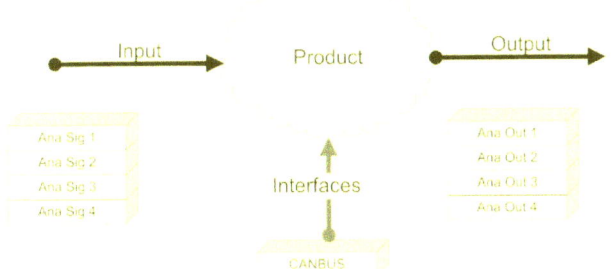

Figure 2.7 Black box visualization

An inventory of properties and behavior is made. Signals that need to be processed or need to be generated are also documented in this stage.

The SPECIFY stage is always concluded with a context diagram that shows a clear border between the outside world and the actual parts that needs to be developed.

In this stage, the documentation of all specifications and requirements must be considered the most important task. If ever, these things are not correctly documented or incomplete, you could have big issues later during the project. Remember that a customer is not automatically an expert on how things are done just because he has a great idea! It is therefore essential for the designer to help the customer with documenting these specifications. It takes experience to properly document the specifications and requirements. However, there are some great tools to help you in this important task.

2.3.1.1. Specifications & requirements Card

Let me give you a good advice to fill up your designer's toolbox with tools that you will be using over and over again. Over time, the experience will grow and the content of this toolbox can be adjusted as needed.

De following illustration hands you the first tool. Feel free to use it during the interview and the documentation of the specifications.

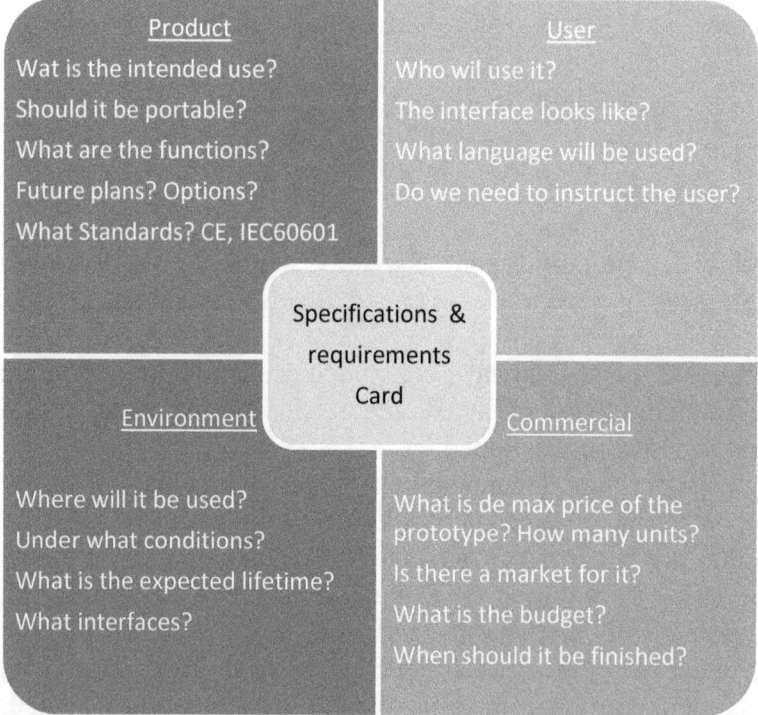

Figure 2.8 Specifications and requirements card

Of course, the tool is far from being complete. It is up to you to complete it! You can print it and use it as a tool. The purpose of this tool is to make the designer think about what might be important for this project. When more experienced, the number of questions will increase. The best advice I can give on how to get all specifications and such on the table, is to keep asking questions!

Another example I can give you is one in which a customer asks a designer to build him a time machine. The initial reaction of the designer was a big smile on his face while he was wondering if he was dealing with a crazy person. Everybody knows that time travel is science fiction, right? However, as a more experienced designer he then asked: "Tell me what this time machine of yours should be capable of." The answer given by the customer was one you might not expect. You see, all he wanted was a machine that shows time. He was asking the designer to build him a clock!

After all specifications and requirements are documented, they should be listed in a table together with some sort of priority.
For this, I like to use the following:
W : Wish
D: Demand

A wish is something that would be nice to have but not having it would not be critical.
A demand however, is critical and not having it would mean that you are dealing with a failure in design.
Sometimes, for whatever the reason might be, it is not possible to realize a demand. It is essential to communicate this with the customer immediately to see if you can agree on a solution.

Specifications/requirements	Wish/Demand	Check
Product		
4 ADC inputs 0.5V_{tt} max 1Khz	D	☐
Max power < 10W	W	☐
Supply 24VDC	D	☐
User		
GUI in Dutch and ENG	W	☐
Ipad interface	W	☐
Environment		
-20°C > Ambient > 40°C	W	☐
Explosion safe	D	☐
CAN bus interface	W	☐
Commercial		
Budget max €8000	D	☐
Prototype finished before dec. 2019	D	☐

Figure 2.9 Specifications and requirements table

When the inventory of specifications and requirements has been made it is important to look at them critically. Are these the right specifications and demands? Are they complete? Does one not rule out another? Are they realistic? Reasonable? Can we validate them?

2.3.1.2. SMART method

A good method of describing the specifications and requirements is the SMART method. The SMART method states that we can only use the name specification or requirement if it complies with all parameters of the SMART formulation. The Smart method has been proven in the field, to be a good way to help you make specifications that are within range of what is possible.

Although it is highly recommended to use all parameters of the SMART method when forming specifications and demands, it is fair to say that sometimes it can be enough to just follow some and not all parameters. As for every other tool described in this book, the use of the tool should help you in any way you see fit. Sometime, using a tool has no extra benefits. It is of course up to you to decide what tool to use and how to apply it.

The SMART parameters are:

- Specific: Target a specific area for improvement. There is no room for interpretation on what is described;

- Measurable: Quantify or suggest an indicator of progress;

- Acceptable: There is enough support to make it happen;

- Realistic: The result can realistically be achieved;

- Time-related: Specify when the result(s) should be achieved;

An example of a definition that is not compliant with the SMART parameters:
The room temperature should increase rapidly whenever the outside temperature is cold.
An example of a statement that is compliant with Smart parameters:

The room temperature should increase from 16 °C to 20 °C in less than 30 minutes whenever the outside temperature is below 2°C.

Finally, we conclude the SPECIFY stage with a context diagram of the product. A context diagram shows the relation between your product and the outside world. In one diagram, it is possible to see what parts belong to the product and what parts do not.
The level of details inside the context diagram is up to you to decide. It is common to include the parts of the user interface because they are connected to the outside world.

Below, you can see an example of a context diagram. More details about context diagrams will be addressed later.

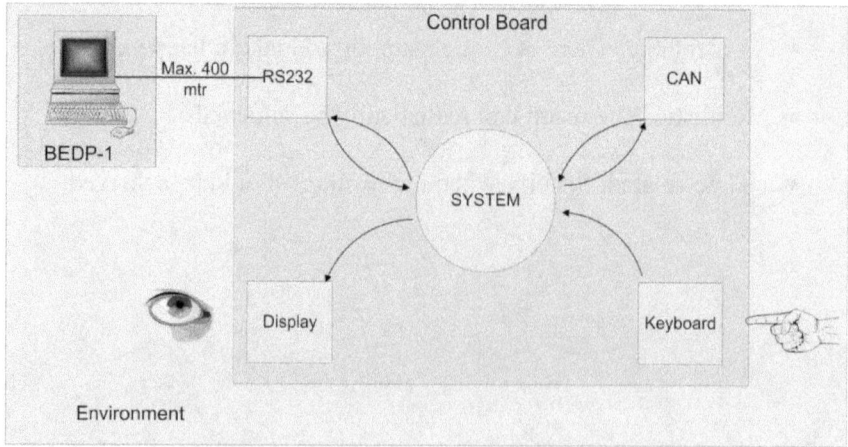

Figure 2.10 Context Diagram

2.3.2. The Design stage

The previous stage has been concluded with a representation of the system as being a black box by the use of a context diagram.

In the DESIGN stage, the context diagram will be split up into small designs. The smaller design will then be described superficially. Also, all solutions will be listed in a morphologic table. We will end this stage be choosing one possible solution for every smaller design. With this selection of solutions, we will then move on to the CREATE stage to work out every little detail.

First, the black box should be divided into a workable number of smaller black boxes.

As is shown below, this collection of smaller boxes is presented as a cloud. Inside the cloud the smaller black boxes are shown.

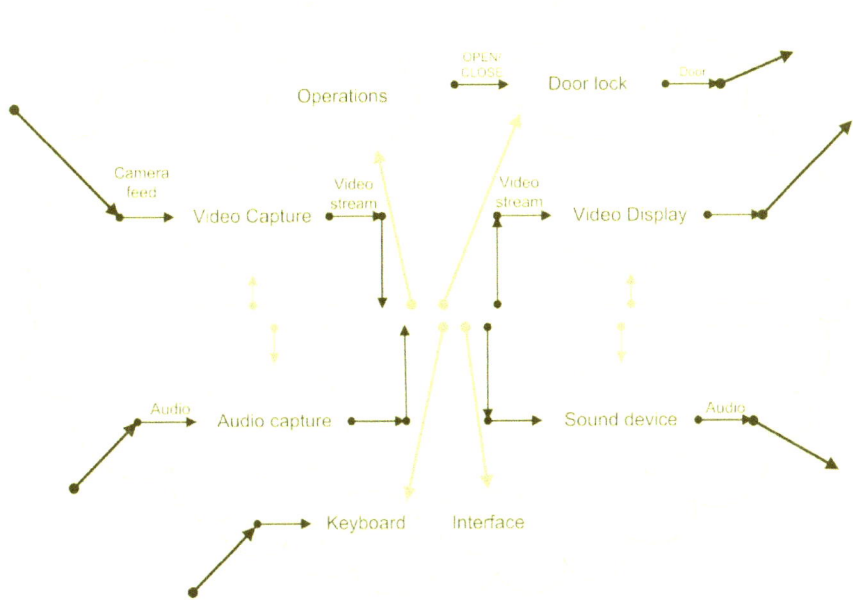

Figure 2.11 Design Cloud of a system

In the middle of the design cloud, as shown on the previous page, there is some sort of bus. This is what is called the lifeline of the system and is often realized later in the design by using a microcontroller. The red arrows represent the control of this black box. (Let us just call them partial designs for now). Every partial design is shown as a little cloud and has its own specific properties.

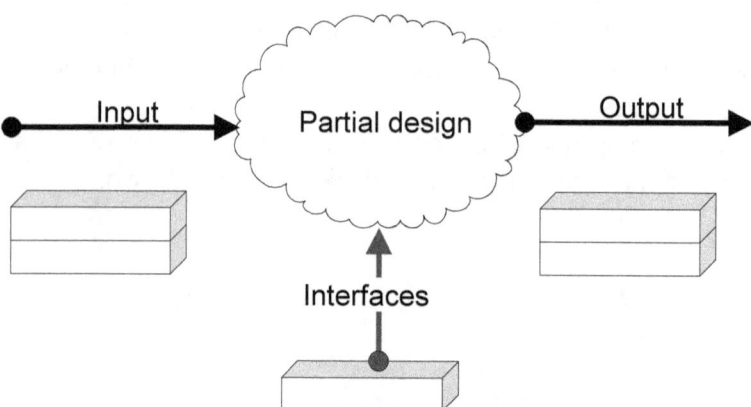

Figure 2.12 Partial design black-box representation

If the partial design has inputs, the inputs will be represented by using an input arrow as shown in the figure above. Outputs are represented in a similar way. We also have an arrow that represents the interface and the name of the interface can be mentioned as well.

Once the system has been divided into partial designs, one or more solutions should be found for each single partial design. Do not go into detail this time but just try to find some sort of direction in which we can find the solution. An example can be found while thinking about how to transform an analog signal into a digital one. This can be realized by using a DAC or a R/2R network or even an ASIC. (To name a few).

Mechanical parts can also be included in this part of the design process. An example of this is a door opener. A few possible solutions are drawn below.

Figure 2.13 Possible solutions (Example)

In the next stage you will work out every detail for one possible solution per partial design. It is not always necessary to think of more than one solution for partial designs. Sometimes, there is instant clarity when you think of a solution. However, it is recommended to consider other options that might cost less or are more reliable.

2.3.2.1. Finding solutions

There are several options to find solutions to your challenge. (By Challenge, I mean the fact that you have to find possible solutions for your design). First of all, a designer should use his personal experience and the experience of the team, colleagues and working environment. They can be a great source of inspiration.

What did we do before and how can we use that now?

Moreover, it can be rewording to look around. How did others do this? What is available on the market? Whenever you start working on a project by yourself or as part of a team let me remind you that there are already many tools available that have proven their worth to many alike. Many tools will be addressed later but let me give you some ideas.

First, there is the well known brainstorm session in which a group of people shares ideas. Another method that is less known to many is the brain writing. Both techniques have the same purpose to stimulate the creativity of the designers. Depending on the number of people involved in a brainstorm session, it can be rather expensive because many people are involved simultaneously and the duration of a session is time consuming.

Brain writing on the other hand is a lot less expensive. A simple brain writing session takes only a few minutes per participant. More of these interesting methods will be described later.

When all partial designs are defined, the designer has to make a choice. This is the time that the designer chooses what possible solutions will be designed in detail during the CREATE stage.

At this point, it is essential to know all prospects of a possible solution. Every designer has a backpack of experience but the content of this backpack is different for everyone. Nevertheless, experience troubles the mind. Someone who has more experience with a certain technology might be prejudiced. A designer's experience and expertise will have a big influence on the choices he will make. This is not necessary a bad thing! Also, it could be useful to pick someone else's brain on these matters. Two heads are better than one.

The toolbox, as described later on in the book, will hand you some tools that can be used during the DESIGN stage.

2.3.2.2. Morphologic Table

If a choice has to be made about what possible solutions to use for a partial design, the use of a morphologic table will prove to be very useful. A Morphologic table shows the designer all the possible solutions for all partial designs in one table. Once a choice has been made, the chosen solutions are connected using a line. That way it is not only possible to see all the solutions, but also the solutions that will be used in the CREATE stage.

Partial Design	1	2	3	4
Audio Sampling	ASIC	Analog circuit	Software	ADC
Video Interface	Hire 3th party	LCD CAM unit	Webcam SFF PC	
Door mechanism				

Table 2.14 Morphologic table

2.3.3. The Create Stage

The DESIGN Stage has been completed and a selection of solutions has been made. During the CREATE stage, these solutions will be designed and completed in every detail.
To refresh the overview of the system in your mind, the CREATE stage starts with showing the context diagram again. Another option is to start from the morphologic table.
The context diagram shows a static link between the product and the environment. In order to get to know the dynamic properties it would be better to make use of Use Case diagrams and use cases.
Use Case s work best if used for the whole system, software and hardware.
The CREATE stage should include all information necessary to complete a design. This could therefore include information about the PCB design or the part list or others. It is up to the designer to decide what to include. Usually, an assembly drawing of the printed circuit board will suffice. Documentation for producing the product is usually kept separately.

2.3.4. The Validate stage

The VALIDATE stage should not be looked at as an isolated stage but more as a lifeline throughout the whole process of designing. It is very important to keep the specifications and requirement of the product in mind at all times. Only then, a designer can tune and adjust the design on time, if necessary.

However, there will be a time to test your design and to check if the design complies with what has been agreed upon.

This is done during the VALIDATE Stage. To validate all, the designer can use the same table that was created during the SPECIFY stage. Now the purpose of the last column will be clear. Only if all the specifications and requirements are checked, the product is ready to be handed over to the customer.

Specifications/requirements	Wish/Demand	Check
Product		
4 ADC inputs $0.5V_{tt}$ max 1Khz	D	☐
Max power < 10W	W	☐
Supply 24VDC	D	☐
User		
GUI in Dutch and ENG	W	☐
Ipad interface	W	☐
environment		
-20°C > Ambient > 40°C	W	☐
Explosion safe	D	☐
CAN bus interface	W	☐
Commercial		
Budget max €8000	D	☐
Prototype finished before dec. 2019	D	☐

Figure 2.15 Specifications and requirements table

It is important to document any specification that is not compliant. No matter what stage in the design process you are in! That is the only way to prevent having the same discussion over and over again and it helps others to understand the decisions that have been made.

Of course, it is easy to look at a table that the designer most likely made by himself and to mark every item in the last column. The question is how others can check that everything is compliant to the specifications. This is exactly why writing testing procedures is necessary. Not all can be tested by following procedures but they should be used whenever possible. Procedures for testing and validating products create trust in the product. Also, if correctly documented, anyone could test a product for you.

The VALIDATE stage is perfect for tightening up your specifications. During the SPECIFY stage some specifications might be open for discussion but during this stage, discussion is no longer an option. Everything has to be clear. Specifications like "a minimum of 4 inputs" can now be better describes: "4 inputs". Or maybe you added some extra: "8 inputs".

If all specification and requirements are met, the product will not be changed any further during this design process. If the customer wants to change something he should place another order for a new project, right?

If the budget allows, you could ask a colleague to look at the product and you could let him try to write down the specifications that are discovered while testing the product. Do not give him the list of specifications you already have. He might look at the product differently and you might find some new interesting specifications.

3
The toolbox

'That's something I never want to do: I never want to think I know it all, because I don't. There are always more people with more advice, and I just want to soak all that up and make my sporting toolbox as full as I can get it.'
Kirsten Sweetland

The toolbox has a large collection of tools and methods that can help a designer in his quest to make a good design. You might not have a use for all tools and of course; a tool should only be used if it helps you in your quest. It should not be a burden to use a tool because that would never help you in any way. Yes, it might take some 'getting used to' using a tool. But, once you mastered its use, I am sure you can make a balanced decision. There is nothing more frustrating than spending a large amount of time and effort in something for wh ich you have no purpose. For example, do not make use of a morphologic table if you only have one solution for every partial design. Also, there is no use in making extended flow charts to describe a door bell. Only use a tool if you have a use for it! In the table below, you can see what tools can be used in what specific design stage.

	Plain text	Interview	Requirements card	Mind map[1]	Protocol diagram	Activity diagram	Brainstorm/writing	Creative Tools	Context diagram	Flowdiagram	State diagram	Nassi Schneiderman	Morphologic table	Bubble diagram	Use Case diagram	Use Case
Specify	•	•	•	•	•	•	•	•								
Design	•			•	•	•	•	•	•	•	•	•	•	•	•	
Create	•			•	•	•	•	•	•	•	•	•			•	•
Validate	•			•				•	•							

Figure 3.1 Toolbox overview

---

[1] Mind Map is a registered trademark of the Buzan Organisation Limited 1990, www.thinkbuzan.com

3.1. Specifications & requirements Card 'SPEC-CARD'

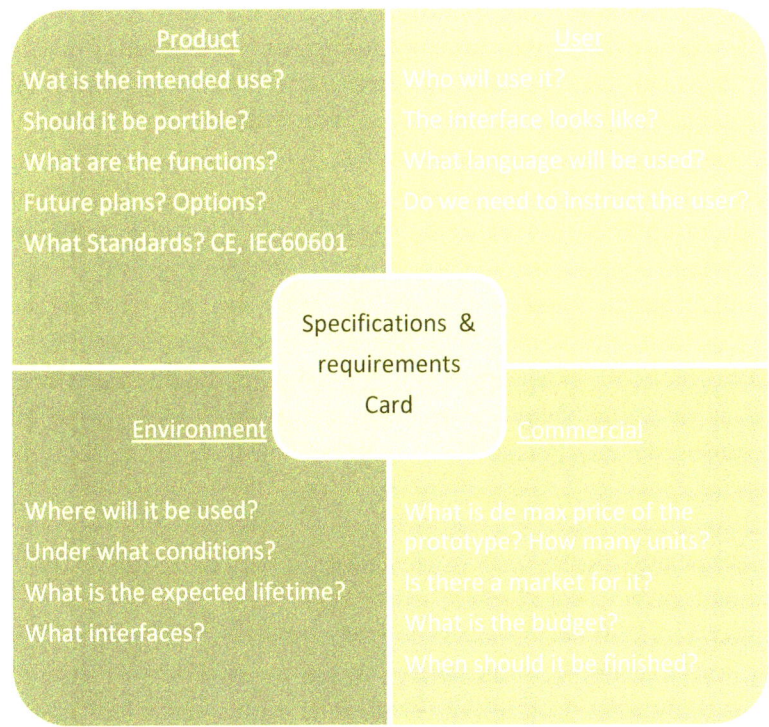

Figure 3.2 Specifications and requirements card (SPEC-CARD)

De 'Specifications and Requirements Card', (SPEC-CARD) can be used to help you make an inventory of de specifications and requirements. The questions that are listed on this card can be changed by the designer. Maybe you like to add some questions? Make this card, **your** card! Change it any way you see fit.

3.2. The interview

The 'first contact' between designer and customer usually extends to a 'follow-up' meeting. In this meeting, let us call it an intake interview, the customer will explain what it is that he want to have designed. This is the right time to talk about specifications and requirements. Do not forget to put them in writing! It is important for the designer to prepare himself for this intake and it really helps if he makes himself familiar with the background of the customer. Who are you dealing with? Is he a technician, an engineer or maybe an entrepreneur? Also, you should know the ins and outs of your company. Make sure you know what you are capable of and clarify beforehand, what you may or may not do within the company to realize this project. What are the lead-times within the productions facility? What are the rates? What technical disciplines are at your disposal? Maybe the customer already told you a bit about what he needs? Perfect! That means you have time to study this theme.

There are many kinds of interviews. Two extremes are a standardized interview and a free interview.
The standardized interview is based around fixed questions that you will find answers to. This kind of interview will get you started but it isn't really that productive for an intake interview. During an intake, the customer should be the one who does most of the talking. The job of the designer during this interview is to get all the facts like specifications and requirements on the table.
As mentioned before, the designer should not attend this meeting unprepared.
During a free interview there is more time for all parties involved to take part in the conversations. During a free interview, the designer can use a list of words as a reminder to trigger questions. Also, the use of the SPEC-CARD is highly recommended.

Beware of the dog! During a free interview you might get lost talking about interesting things that are not really helping you in getting the facts on the table. A designer that is not very experienced might spend a lot of time talking and listening without getting his questions answered. Sometimes it is necessary to do a second or a third interview. Also, you should be aware of possible misinterpretation. There is only one advice I can give you in this regard; keep asking questions!

3.2.1. Preparations

- Be prepared and make sure you know who you will be talking to. What is his educational background? What is his job?

- Make a list of word, topics that you want to talk about.

- Study the 'SPEC-CARD' or if possible, bring it.

- Don't forget to make an appointment.

- Make sure you know what you and your team are capable of and within what time span you can get things done.

3.2.2. The interview

- Make sure you are able to have a conversation in a language you both can understand. Not all customers speak your native tongue.

- Ask clear and firm questions and make sure the other person understands what it is you are asking

- Explain why you are asking a question if necessary.

- Be critical and do not be pleased with a given answer too quickly.

- Listen carefully to the answers and give feedback. ("So, do I understand correctly that you want…..")

- Ask more questions if necessary.

- Take notes!

- Try to take and keep the lead in the conversation. It is your job to get all the specifications and requirements on the table.

3.2.3. Documenting specifications and requirements

It speaks for itself that the designer takes notes during the interview. It's the designer's task to translate these notes into specifications and requirements. If something is not clear, whatever the reason may be, make sure you clarify things before they're finalized. Facts are facts and assumptions are just assumptions. If you need to assume something to complete a specification, make sure you also document this. Remember that assumptions are not always correct and incorrect assumptions may get you in trouble at some point.

You should prevent things from being unclear and you should tackle them before they raise their ugly head.

3.3. Mind mapping

Mind mapping is a commonly known tool that can be used to organize information. It can be used during the interview to quickly note specifications and demands. It can also be used to support the creative thinking during the DESIGN stage.

The mind map is centered on a topic like a picture or text.
Starting from the topic, different branches sprout to show things that are related to the main topic. Sometimes, the use of small pictures or icons can keep the mind map readable.

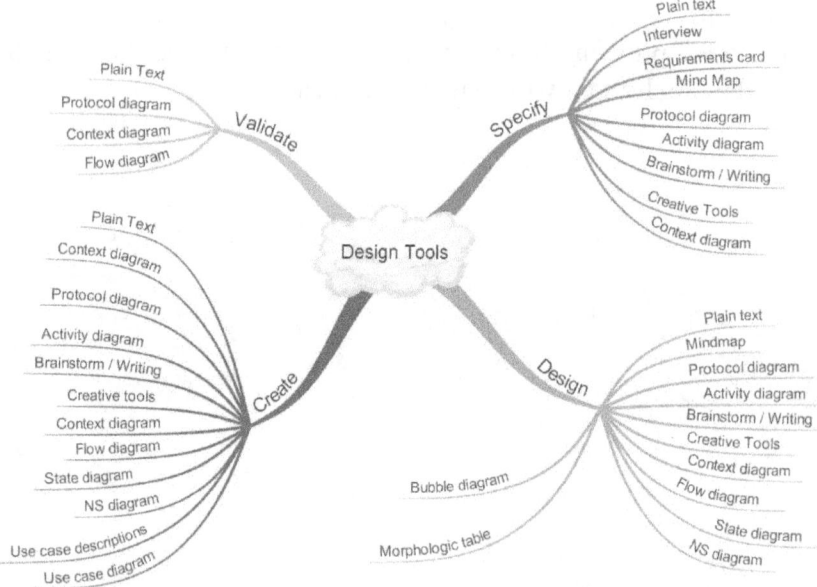

Figure 3.3 Example of mind map for product specifications
Created with iMindMap www.ThinkBuzan.com.

3.4. Use Case diagram

A Use Case Diagram is part of the UML language (Unified modeling language) and can be used to visualize the design's static behavior.
A Use Case Diagram is most useful during the SPECIFY Stage.

A Use Case describes the different functions of the partial designs. Since the main design will be split up into partial designs, there will be an equal amount of Use Case diagrams. The Use Case diagrams will also show actuators. In many cases, a person will take the role of an actuator who has something to offer or is in need of something but is might as well be actuators or sensors. In the figure below there is a Use Case Diagram of a small car. This little car is used to scout a field after the controller presses the start button. Whenever the car encounters an obstacle, the car will stop and an alarm will sound.

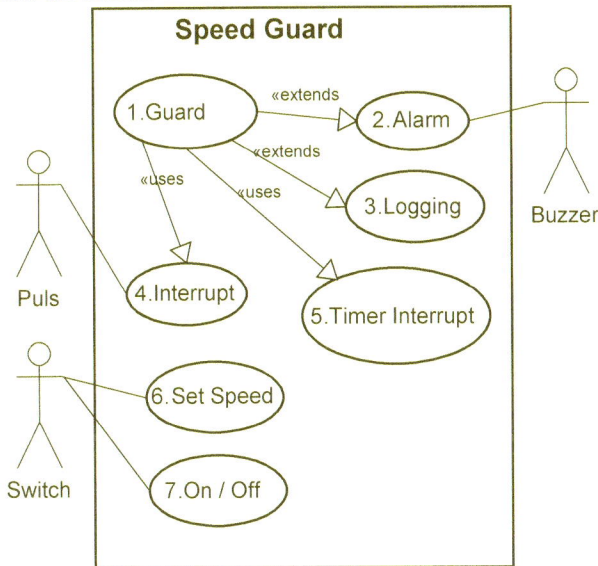

Figure 3.4 Example of a Use Case diagram

3.5. Use Case Description

Use Cases describe the dynamic behavior of a system. It is very common to combine the Use Case Diagram (static) with the Use Case Descriptions. (dynamic)

Use Case Descriptions	
Name	A unique name of the use case
Summary	A brief description of the Use Case function.
Actors	A list of all actors (actuators) that have a relation with the use case.
Related to	A list of other use cases that have a relation with this use case.
Assumptions	Assumptions that have been made when executing this Use Case
Description	A step by step listing that describes the functionality of this use case.
Exceptions	A list of exceptions for which this Use Case will behave differently.
Results	A list of result. What did the Use Case achieve when it is finished?

Figure 3.5 Use Case Descriptions

3.6. Bubble diagram

A bubble diagram can be used to map the relation between partial designs. Embedded systems are often built around a microcontroller and all partial designs can relate to the microcontroller.
A bubble diagram can also be used to map the relations between functions that are within the same firmware. It can even be used to map the relations between functions in software and partial designs. In other words, the internal relations of the complete design can be mapped using a bubble diagram.

A bubble diagram uses a circle to represent a function or partial design. This is the so called bubble. Hence the name bubble diagram. Next, the relations between the different bubbles are represented with lines. These lines can have different colors and its thickness can differ. It is also allowed to use interrupted lines. Every line represents some sort of relation. Some examples are:

- Wireless connection (Interrupted line).

- Direct hardwired connection (uninterrupted line)

- Software connections(Double line)

The next figure will show an example of a bubble diagram.

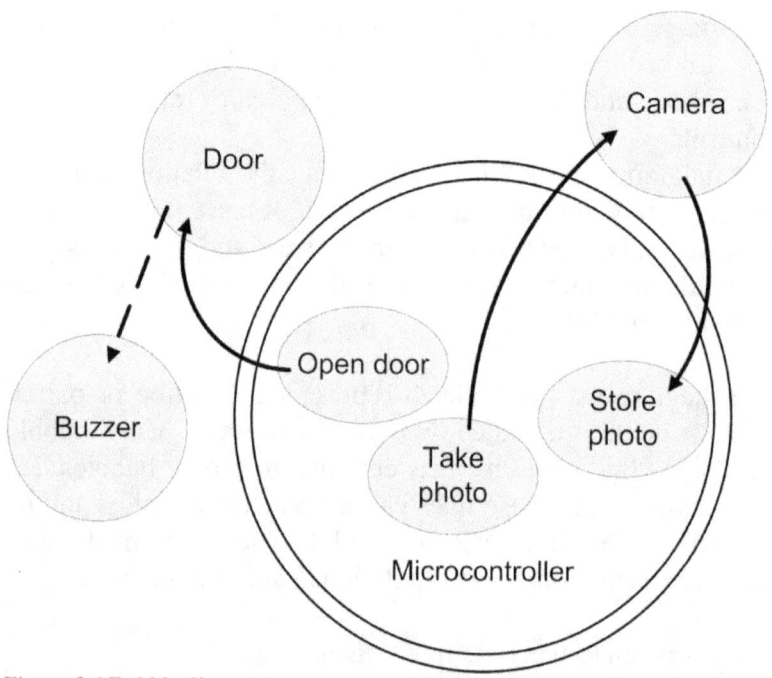

Figure 3.6 Bubble diagram

The figure shows a bubble diagram of a door spy. Different relations between the partial designs can be seen.

3.7. Context diagram

By using a context diagram it is possible to visualize the product in the environment. It shows exactly what part belongs to the system and what part to the environment. It is up to the designer to decide about the level of detail that is used in the diagram. Usually, all parts of the user interface are included because they relate to the environment.

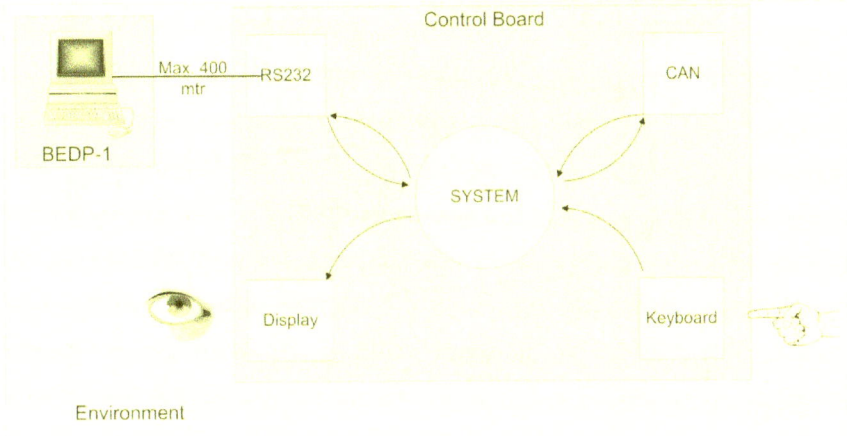

Figure 3.7 Context diagram

The figure above shows an example of a context diagram. In this example, the designer chose to illustrate the most important parts of the system. A sort of interaction is recognizable within the environment. If a design is made up of several partial designs, the designer can use several object in the diagram. In the figure above, this is done by using several blocks that communicate with each other. There are several versions of context diagrams. Some versions use arrows to give direction to communication lines. This is also done in the example above. There are also simpler versions available that do not use any interval arrows. The context diagram is used to illustrate the system's static behavior.

3.8. Protocol diagram

A protocol diagram can be used to map protocols or ways of acting that are arranged in their time sequence.

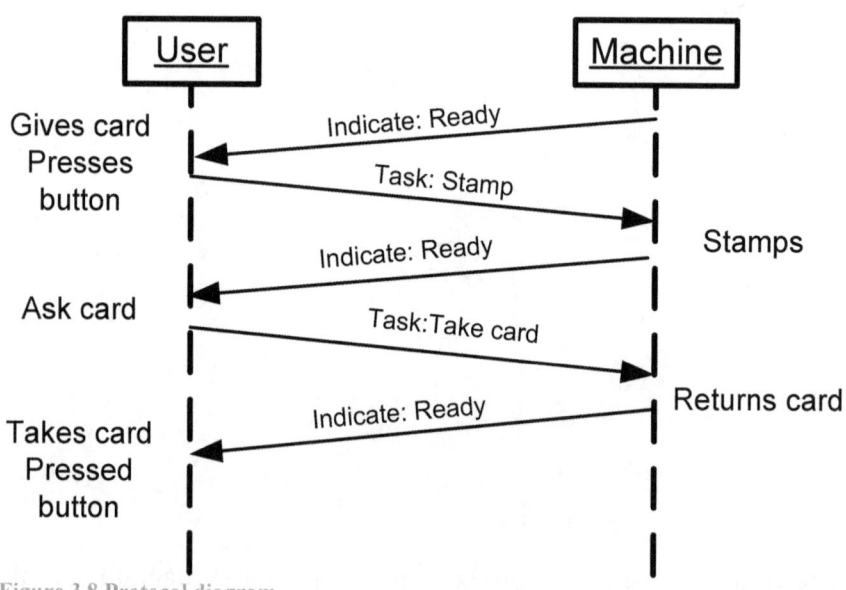

Figure 3.8 Protocol diagram

In the protocol diagram all actors are represented by a vertical dotted line with their nametag on top. In between these vertical lines all kind of actions are illustrated that happen between them.

A similar diagram can be used to indicate the different parameters between software functions but there are better tools available for that purpose. (UML)

3.9. Activity diagram

By using an activity diagram it is possible to illustrate the different processes within time and their relation toward each other. This tool is especially useful for RTOS systems because they use several processes that run simultaneously. (real time operating systems).
The illustration below shows an example of an activity diagram that was made for an imaginary carwash.

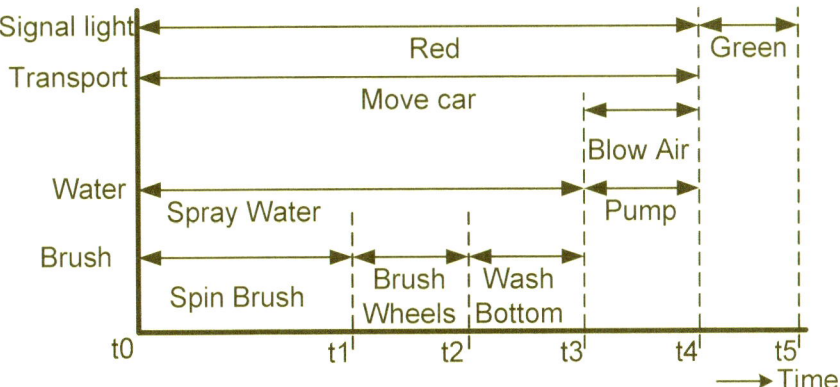

Figure 3.9 Protocol diagram

On the vertical plane the different actors are represented. On the horizontal plane, the different actions are written as a timed function. So called moments of synchronization can be easily distinguished by following the arrows.

3.10. Flow diagram

The flow diagram is a well-known tool within all kinds of disciplines. They are used to illustrate the order of how things happen. This can be about software, hardware or even surveys and communications in general.
A funny example is shown below:

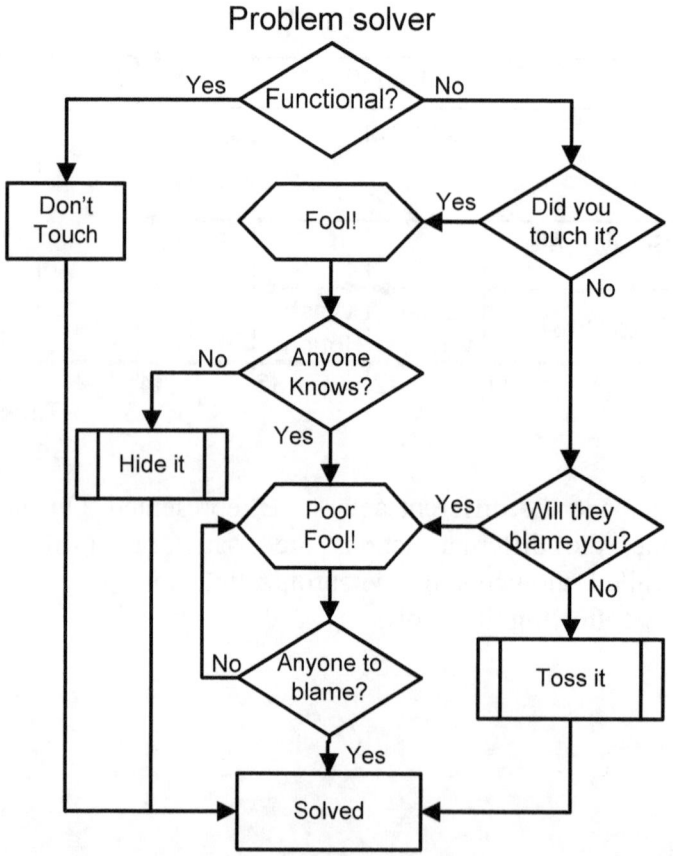

Figure 3.10 Flow diagram example

The shapes that are used in the flow diagram are various. Because there are many versions available, a standard was created. (In this example the standard is not used.)
Most important about a flow diagram is that it should be readable to the intended readers. It really doesn't matter that much what kind of shapes you use.
If you want to know more about standard flow charts, please take a look at the standard, (for example: ISO 5807:1985 and ISO 10628) A few commonly used shapes are listed below:

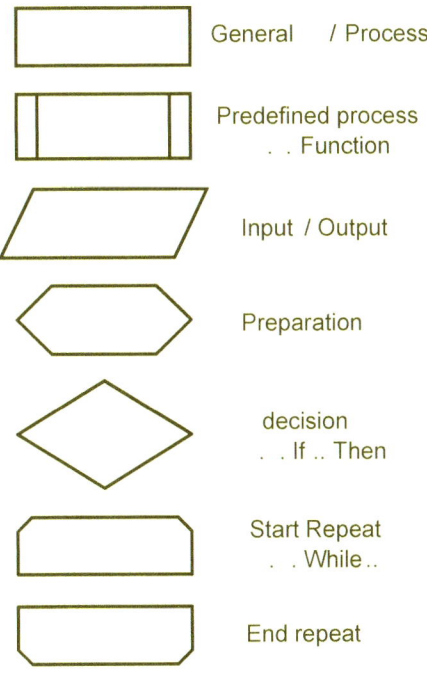

Figure 3.11

3.11. Nassie Schneidermann diagram

A variation to the flow diagram is a Nassie Scheidermann diagram. (NS-diagram in short) It is also known as PSD (program structured diagram). A NS-diagram is no more than a schematic representation that illustrates the way that instructions co-relate.
Basically, a NS-diagram can be used for the same things a flow diagram can be used for. The biggest advantage is that the NS-diagram is more compact than a flow diagram.

NS-diagrams contain:

- Text

- Rectangles

- Triangles

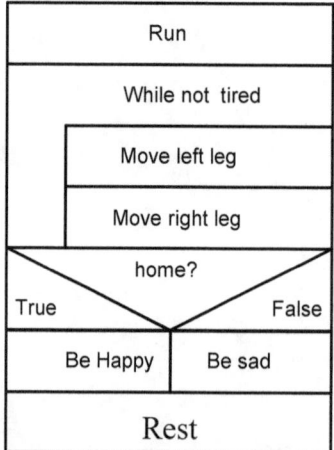
Figure 3.12

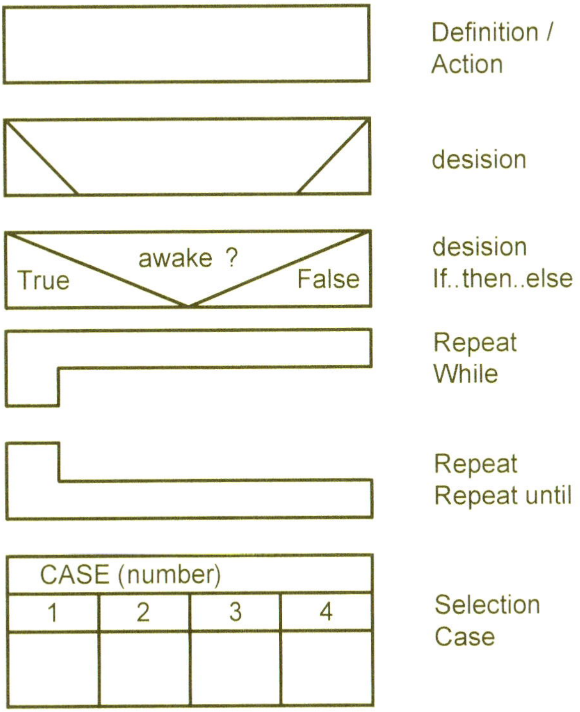

	Definition / Action
	desision
True awake ? False	desision If..then..else
	Repeat While
	Repeat Repeat until

CASE (number)				
1	2	3	4	Selection Case

Figure 3.13

3.12. State diagram

A state diagram can be used to illustrate the different states a system or partial design could be in. State diagrams can therefore be used to enlighten the dynamic behavior graphically.

Many versions of state diagrams are available but they basically work the same. Some versions use circles to represent a state while others prefer to use squares. The shape really doesn't matter that much. It's all about making a readable illustration for the intended reader.

3.12.1. Finite state machine

Finite state machines use the following terms:

- A system has a limited number of states

- A state of a system is determined by the value of one or more of its attributes.

- Triggered by external events, the system can move from one state to another. (Conditions)

First, the number of possible states are determined and named.

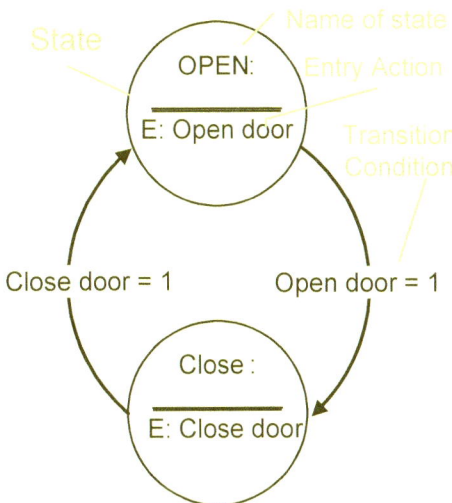

Figure 3.14

Next, an inventory of all possible actions within the state is made. Upon entry of another state, an action can take place. These are the entry actions. The same goes for exiting a state. (Exit actions)

Activities can be interrupted. Actions however, cannot be interrupted. Actions will always finish. To move from one state to another, one or more conditions have to be met. In some versions it is allowed to combine actions with conditions.

Depending on the number of states a system could be in, different colors or types of lines can be used to illustrate the states. That really helps to keep things readable when the number of states is enormous. Some examples of this are listed below:

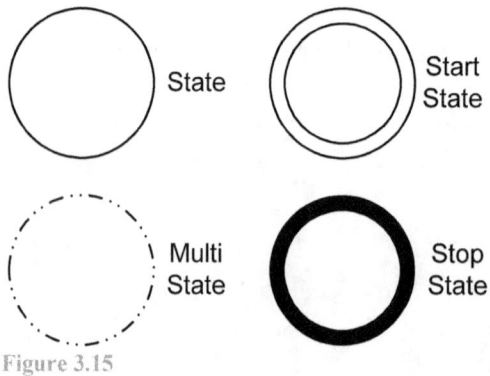

Figure 3.15

The method as described above is rather simple and therefore somewhat limited in possibilities:

- No hierarchy and therefore hard to read.

- No concurrency of tasks and therefore not usable for RTOS

3.12.2. State chart

State charts can do everything that finite state machines can do combined with a little extra:

- Hierarchy is possible by using super states.

- Concurrency is possible

A super state is a state machine that has its own internal state machine. The change from one state to another happens the same way as for the finite state machines, using transitions.
The change from a super state to another (super) state happens with a group transition. A group transition always has priority over a normal transition. The process of the internal state machine of a super state will be interrupted if a group transition takes place.
Entry and exit actions can be used for normal states and super states. A state machine always starts from a starting point, represented by a dot. The same goes for a super state. The internal state machine of a super state always begins from its own starting point, also represented by a dot.
The next figure shows an example of a state chart:

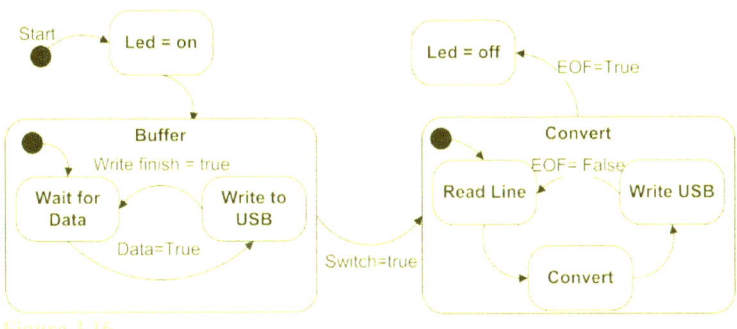

Figure 3.16

Concurrency can be illustrated by the use of a dotted line. The super state in the next example shows two processes that are running simultaneously:

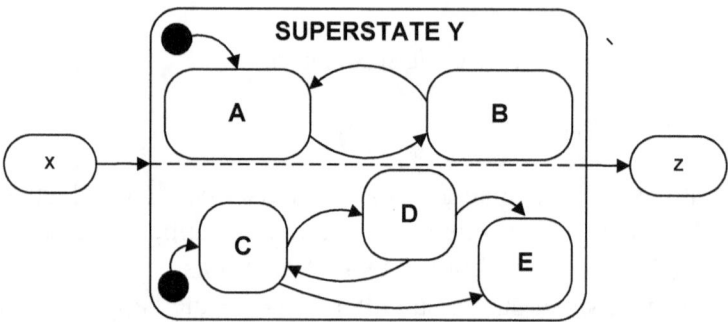

Figure 3.17

3.13. Brainstorming & Brain writing (wiki)

Brainstorming is a way to stimulate creativity during a meeting. The purpose of brainstorming is to quickly create a lot of new ideas about one or more topics. The most important rule of brainstorming is that no idea will be criticized during the inventory of ideas. People have to keep an open mind. Criticism can follow at a later time.

The rules of brainstorming:

- No criticism towards ideas that are put on the table. Criticism will follow at a later time.

- The quantity should be kept at a high level. The idea behind brainstorming is to pick each other's brain to get as many ideas as possible on the table.

- Wild or strange ideas are welcomed with open arms. Sometimes unexpected ideas lead to great products.

- By combining wild ideas with existing ones, great things might be achieved.

- Try to follow a strict time or goal and stick to it.

The process of brainstorming knows 3 stages:

- Prepare
- Generate
- Evaluate

The prepare stage.
A group of people is formed. This group should contain specialists as well as some people with expertise unrelated to the topic at hand. These so called 'wild geese' have a different perspective on things and might come up with some surprising ideas.
The group will then be informed about the problem at hand and the brainstorming can start. There is a chairman who will direct the brainstorming process.

Generate stage.
The chairman collects all the ideas. This is usually done on a white board or something similar. The chairman makes sure that everybody gets their say and he makes sure that no ideas are criticized. This stage can take as long as the chairman sees fit but it is always good to set a time limit that will not be exceeded.
It is important that the setting is relax and creative in order to be productive.

Evaluate stage.
After all ideas have been collected they will be evaluated. This can be done by the team or another team of specialists.

Brain writing
Compared to the time consuming brainstorming, brain writing is a quick and easy process to generate a bunch of ideas in a short period of time. Both methods are similar except that brain writing is more private because it can be done by email if necessary.
But to explain the process, let us just assume that everybody is sitting in the same room. The chairman will give everyone a piece of paper. The paper shows an explanation or description of the problem at hand and the chairman will ask everybody to write down one possible solution. After some time the papers will be handed over to the person next to you. This person will then add another solution to the one on the paper in front of him.

A well known brain writing method is the 5-3-5 method in which 5 persons write down 3 solutions after which the papers will be passed on to the next person 5 more times. That way it might be possible to generate up to 90 ideas for the problem at hand in a short period of time.

The biggest advantage of brain writing compared to brainstorming is that the impact on resources is less for brain writing than it is for brainstorming. If the brain writing is done by email, it will be even less time consuming and it suddenly becomes affordable.

3.14. Creative tools

Everyone can run out of ideas and it can be very frustrating if it happens to you. But if it does happen, please do not give up. On the contrary, this would be the moment to start looking at the problems in a different way.

If for whatever reason, it happens that you run out of ideas, try using some of the creative tools. You have to start thinking 'out of the box'.

A method that is successful and funny at the same time is the method of lateral thinking. It is not only a funny way to create interaction between the team members but it also creates ideas that nobody thought of before.

With lateral thinking all obstacles are ignored to find new or other angles to look at problems.

An example of a problem that can be solved with lateral thinking is de invention of a time machine. Mister Jules Verne believed that traveling thru time is something he would like to do and so he asks a team of developers to build him a time machine.

The modern scientists sometimes disagree on this topic but according to Albert Einstein, time traveling is not possible.

Indeed, that is a limitation that cannot be ignored. Or can it? In the method of lateral thinking we would give this question a twist. Like if we could travel thru time, what would we do? What would we see? How will we talk to the people we meet? In the end, the team of designers came up with something that is now known as an interactive history book or encyclopedia. And they only had this brilliant idea because they looked beyond the limitation by thinking lateral.

Another way of lateral thinking is that you imagine being someone else. If I was a bird, how would I land on my belly? The answer to this question could eventually lead to the design of an emergency landing gear for airplanes. How would the animals do this?

With lateral thinking, there are no limits and you can go wherever your fantasy will take you. Problems and obstacles are ignored. The ultimate target is to stimulate the designers to find different angels to solve a problem.

Another way of thinking lateral is to change the format of the question into one that starts with: "Would it not be great if...."
Would it not be great if there was a book that shows me all the steps for making a great design?

Would it not be fantastic if we could travel thru time? If we could relive the past, today. If we could prevent mistakes in the future by learning from the past?....... Documentation is important after all...get it?

3.15. Morphologic table

If you have to choose between several solutions for the same problem, a morphologic table can come in handy. A morphologic table shows all possible solutions to all problems within the project. Once the right solutions has been selected, they can be connected with a line so that it is clear to see for everyone. You can use different kind of lines to create alternative paths. A morphologic table will not only show you the chosen solutions but it will also show the alternatives. This can be of later use.

Partial Design	1	2	3	4
Audio Sampling	ASIC	Analog circuit	Software	ADC
Video Interface	Hire a 3th party	LCD CAM unit	Webcam SFF PC	
Door mechanism				

Figure 3.18 example of a morphologic table

4
Case study 1: Sports speed guard

'If you are worried about falling off the bike, you'd never get on.'
Lance Armstrong

In this case study, the whole process of the 4-step method for designing, will be used to develop a pseudo speed guard for biking. The result will be a working prototype (on paper only, but theoretically you could build it for real)

4.1. Specify Stage

As with all development that is done with the 4-step method, we will start preparing for an interview, followed by doing the interview. After the interview, we will list down the specifications and requirement using a product sheet. We will also include a context diagram.

4.1.1. Preparations

Your customer, who happens to be a bike store owner, has given you a heads up about his plan to put a bike speed guard on the marker.
So far he has only told you that this speed guard is a device that can be used to monitor speed. He wants to talk to you about his plans and he would like to discuss the possibilities.
With this information at hand, the reason for setting up a meeting is within grasp. The first thing to do would be to set a date for this meeting. During the interview you will represent the designer and the future business partner of this new customer.

- Make sure that you use up to date information about you and the company you represent.
- Make sure you are informed about actual lead time within the company and know what the company is capable of doing.
- Do some research about how speed on a bike can be measured and maybe you can find out what is available on the market.
- Make sure what promises and commitments you are allowed to make on behave of your company.
- Read the SPEC-CARD and try to think ahead about what questions might help you during the interview. Some examples:
    - What exactly is it that the customer want and what is it that he wants you to do?
    - What functionality should be implemented in the device?
    - Can it be done and can it be made at a reasonable price?
    - When should it be finished?

Some questions are part of the preparations while others should be asked during the interview. You are not limited to the questions you think of during the preparations. Look at them as no more than a guide to get you started. Many questions will pop-up during the interview.

4.1.2. The interview

By asking the right questions during the interview, the intension of the customer was clarified.
He is in need of a device that can be used to measure and guard the actual speed during cycling. Operating the device should be easy to understand for all bikers.
The device's functionality is the following:
As soon as the biker presses a button, the actual speed at that very moment is stored. The biker will continue to cycle and the bike will maintain its speed. Although that is what the biker is trying to achieve. Whenever the variation in speed is more than 5% from the stored value, a short acoustic alarm should notice the biker. If the variation is more than 10% from the stored value, a second alarm should sound. The instant the variation is less than 5%, the alarm is reset and the monitoring continues.
With this information, the designer had a better idea about the device but the information was still incomplete.
To get all specifications and requirements on the table, the designer had to ask more questions. So, he used the SPEC-CARD to get started. The questions and answers are listed below:

- Where will the device be used? It will be attached to the bike and it will be used under various weather conditions and temperatures between 0 °C and 40 °C. The device should not be affected by sun or rain.

- Does it have to be portable? Will it be part of the bike? The device should be compact and it will be attached to the steering wheel. It should be detachable and it runs on battery.

- What functionalities does it have? Speed guard: During biking, the speed is monitored and 'locked' when the bikers presses a button. The idea is that the biker has to keep a steady speed to prevent the alarm to give out a signal. The prototype does not need to have a build-in charger but it can be considered optional.

- Future plans? Options? If the product turns out to be marketable, another version with data recording might be developed in the future.

- What standards are applicable? For the first prototype, standards are not really that important but a thorough study of applicable standards should be done before the device is sold to customers. Nevertheless, the device should be safe to use and the developer should consider making the design CE compatible.

- Who is going to use it? The prototype will be used by a group of 4 people to do a pilot study. Future products will be sold and used by consumers.

- What does the user interface look like? A single button to set the speed and a power switch will do the job. The power switch could be replaced by the same button that sets the speed. (double function) The device is equipped with a 'speaker' to give sound.

- What are you aiming for in regards to the maximum productions costs? The customers claims to have a certain budget but he doesn't want to elaborate on that at this time. Instead, he would love to get a quotation from the developer. He will then decide if the project will be granted or not.

- How many prototype units? 3 prototypes

- Is it marketable? The customer believes that it is and he claims to have a potential group of buyers.

- When should the prototypes be finished? 5 months after the project has been granted.

Now, the designer decided to have enough information to make an appropriate offer. Before he can make this offer, all relevant information from the interview has to be converted to specifications and requirements. (The actual offer will not be included in this hypothetical example and we assume that the project will be granted.)

4.1.3. Info conversion

At this stage, it is important to name the product before continuing. This does not have to be the definite name of the product but it is nearly a unique name for identification purposes to be used during development.
The product in this example will get the name 'SPEEDCON'.
At this point, the developer will convert the information he got during the interview to a list of specifications and requirements. During this conversion, the developer had some more questions that he got answered by using email. Questions like how long the battery should less during a ride.
During the interview, it is important to get a complete idea of what the customer wants and by all means, it is allowed for the designer to ask more questions, even after the interview has finished. However, it is wise to collect your questions first and ask them all at the same time rather than calling your customer for every little thing. In general, the customer will always appreciate if you ask him questions about the project because it shows your commitment.

4.1.4. Product sheet

It can be useful to make a graphic representation (black box representation) of the product that needs to be developed. Together with the list of specification and requirements, it creates a good overview of the task at hand and it can be shown to the customer to get some feedback. You can even include this product sheet with your offer and your customer can sign the document to let you know that he agrees with the shown information.

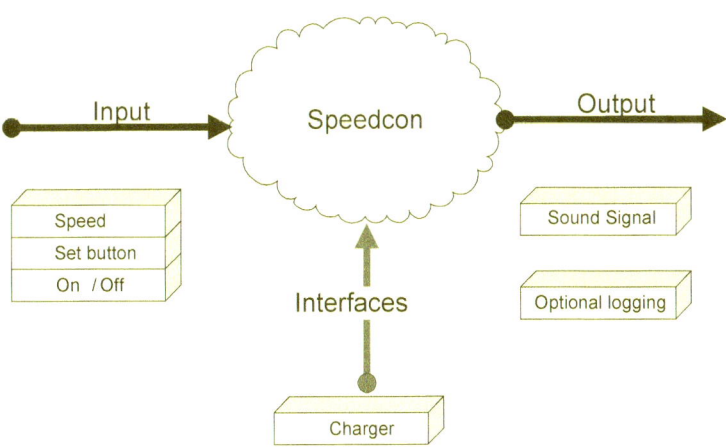

Figure 4.1

Function: To guard the speed of biking, as set by the biker. A variation in speed by 5% will sound two short alarm tones. A variation of more than 10% will sound a long alarm tone.

Specifications and requirements	Wish / Demand	Check
General		
Operational temperature 0 °C - 40 °C	D	☐
Powered by battery	D	☐
Rechargeable	W	☐
Internal charger	W	☐
Time of operation fully charged > 8 Hours	D	☐
Time of operation fully charged > 24 Hours	W	☐
Mechanical		
Water resistant / Splash proof	D	☐
Attachable to steering wheel	D	☐
Future options		
Data recording	W	☐
CE compatible design	W	☐
Functional		
Button to set speed 'set point'	D	☐
On / off switch	D	☐
2 different alarms when speed varies	D	☐
Double function for button to eliminate on / off switch	W	☐
Logistics		
Lead time prototype < 5 months after project start	D	☐

Figure 4.2 This table concludes the list of requirements and specifications.

4.1.5. Context diagram

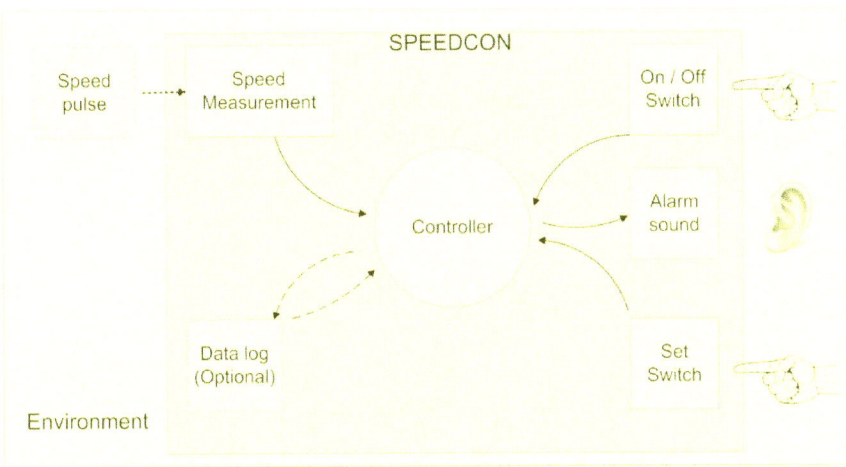

Figure 4.3 Context diagram of the Sports Speed Guard

The SPECIFY stage is concluded with a context diagram of the system. The difference between the context diagram and the black box is that the black box will not show any details at all but nearly gives you an idea about inputs and outputs.

Because the product that has to be designed is rather compact and not to complex, the context diagram includes functional details. When making a context diagram of a more complex design, keeping the context diagram readable might take some practice.

The context diagram of this example shows an individual object on the left side. At this time we do not know how this speed signal will be available. This could be a GPX signal for example, but it might as well be a simple pulse coming from a HALL sensor. On the right site we can see the actors that are part of the environment.

4.2. Design Stage

The SPECIFY stage has been concluded and at this point we know exactly what it is we have to design but we do not know how to realize it at this point. That is why we are going to develop it. During the DESIGN stage several possible solution will be described and we will finalize which one of those solutions will be used in the CREATE stage.

4.2.1. Defining the system

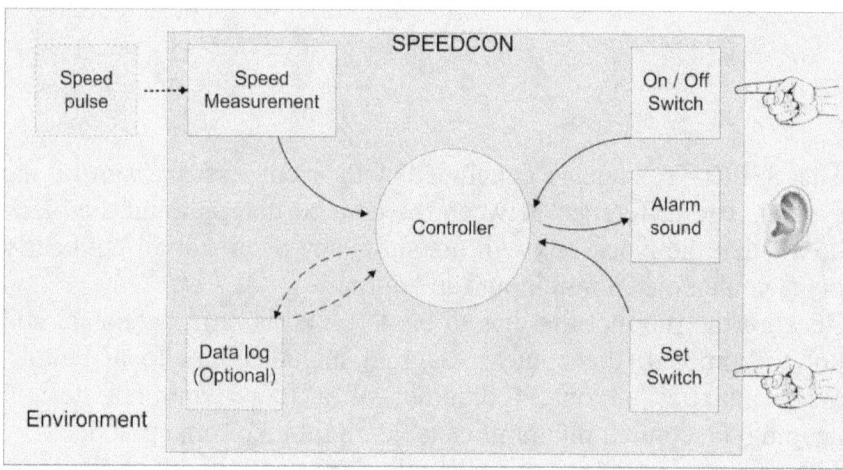

Figure 4.4 Context diagram of the Sports Speed Guard

The context diagram above shows us the system named SPEEDCON. The user interface is clearly visible on the right-hand side and you can see that the user interface is directly related to the user who is part of the environment. The user interface for SPEEDCON is nothing more than a button, a switch and a buzzer of some sort.

Below, you can see the black box representation of this system:

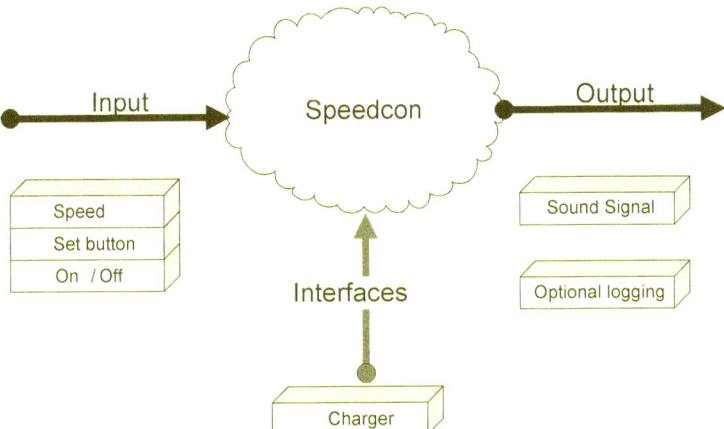

Figure 4.5 black box representation of the Sports Speed Guard system

Function: To guard the speed of biking, as set by the biker. A variation in speed by 5% will sound two short alarm tones. A variation of more than 10% will sound a long alarm tone.

4.2.2. Splitting up into partial designs

The black box of this system has to be divided into several partial designs. For now the following partial designs have been chosen:

- Power Supply
- Optional loader
- Signal transducer / converter
- Microcontroller
- User interface
- Audible alarm
- Optional data logging

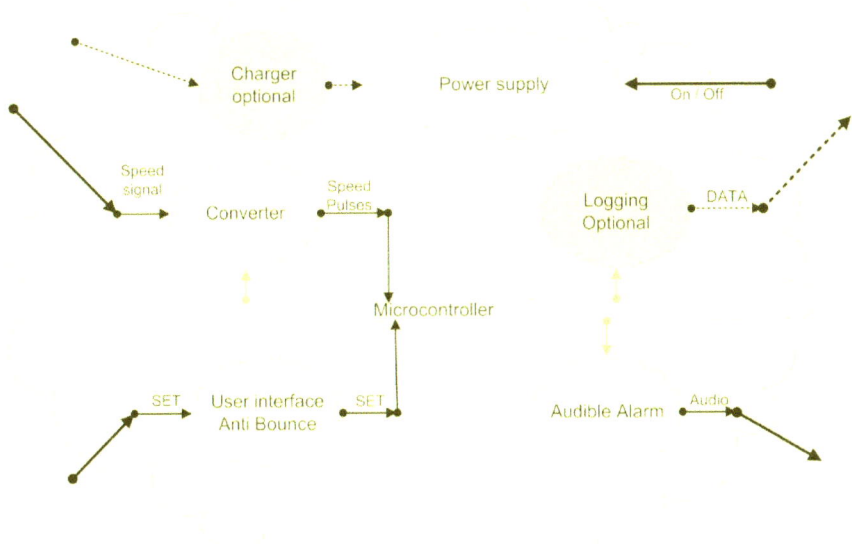

Figure 4.6 Design Cloud of the system

The illustration above represents the 'Design cloud' of the SPEEDCON system and it shows the partial designs that we have defined previously. It is up to the designer to decide if the optional partial designs will be included in the rest of the design process. Costs, risks and development time as well as strategic, all have a part in that decision.

4.2.3. Defining the partial designs

Now it is time to work out some ideas for the partial designs. Let us start by illustrating every single partial design and its properties. Details are not yet available because we have not finalized any ideas at this time. We will use a black box cloud representation for each partial design. (let us assume that the mentioned possible solutions are the result of a brainstorm session)

4.2.3.1. Power supply

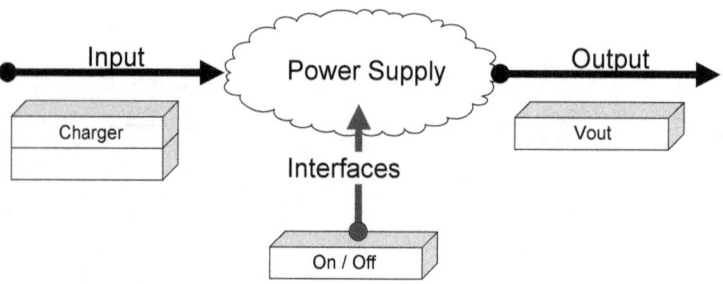

Figure 4.7 Power Supply

Possible solutions
- Batteries
- Internal Power pack (Rechargeable)
- External Power adapter (Home trainer use only)

Solutions	Pro's	Con's
Batteries	Cheap Exchangeable	Limited use with 1 charge
Internal Power pack	Rechargeable More compact	More expensive
Power adapter	Fixed power source Less space consuming	Not possible for normal biking.

Table 4.8 Possible solutions

4.2.3.2. Optional charger circuitry (only when using internal power pack)

Figure 4.9 Optional charger circuitry

Possible solutions
- External power source
- Solar cells
- Dynamo(generator) on bike
- Internal circuit for loading

Solution	Pro's	Con's
Ext. power source	OEM product	Might get lost
Solar cells	No need for external power source	Low efficiency. Large.
Dynamo	No need for external power source Charging only when biking	More electronics needed No standard
Internal circuit for loading	Included so it is always available	More electronics needed

Figure 4.10 Possible solutions

4.2.3.3. Pulse detection / converter

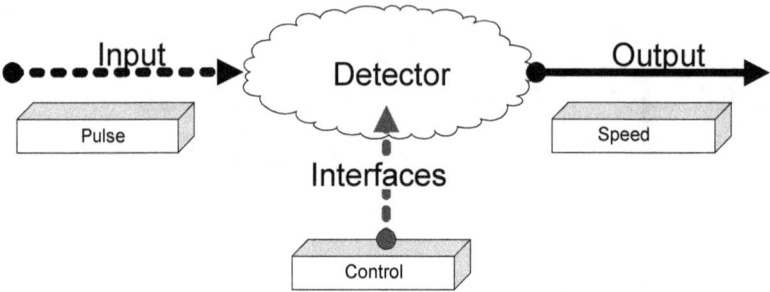

Figure 4.11 Pulse detection

Possible solutions:
- GPS module
- GSM GPRS module
- Pulse transducer on bike
- Reed sensor; Hall sensor; Optocoupler

Solution	Pro's	Con's
GPS module	OEM product	Expensive
GSM GPRS	OEM product	Expensive Less reliable
Reed sensor	Cheap	Limited speed detection for high speed.
Hall Sensor	High speed detection possible	In need of power supply.
Optocoupler	-	Sensitive to dirt. In need of power supply.

Figure 4.12 Possible solutions

4.2.3.4. Microcontroller

The choice of what microcontroller to use depends on its possibilities, costs, operational conditions and last but not least, the personal choice of the developer. (To name just a few)

For this design, the designer chose a controller from Atmel (As you will find out later.)

It is recommended to use similar processors throughout your project in order to save on development time. Also, different processors require different and often expensive tools.

Solution	Pro's	Con's
PIC	Familiar to the designer	Limited tools in house available
ATMEL AVR	Familiar to the designer	All tools available in house
Siemens XL	Cheap USB interface	No tools available in house Unknown to the designer

Figure 4.13 Possible solutions

4.2.3.5. User interface
The control is limited to a single button

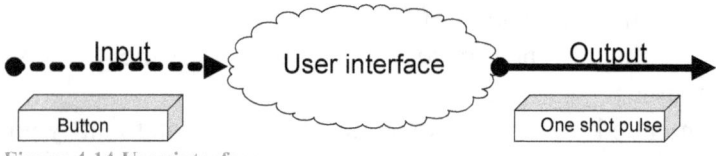

Figure 4.14 User interface

Possible solutions:
- One shot Flip flop (hardware) to prevent bouncing
- One shot anti bounce using a RC network, combined with firmware.
- Firmware solutions.

Solution	Pro's	Con's
One shot FF	Reliable	Negligible
HW & SW	Reliable	Negligible
Software	Reliable	Negligible

Figure 4.15 Possible solutions

Optional control for powering up / down the device:
(Software programming for detecting long button press)

	Con's	Pro's
Software Detecting	No extra power switch needed	Negligible

Figure 4.16 Options

4.2.3.6. Audible Alarm

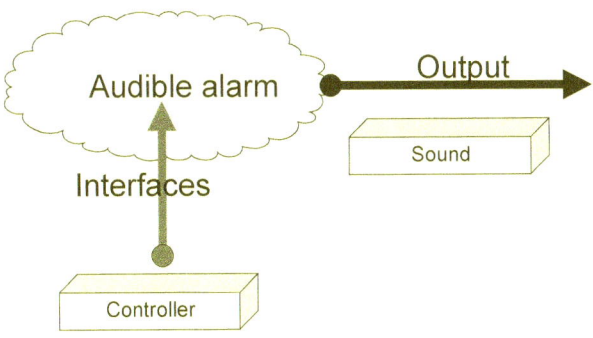

Figure 4.17 Audible Alarm

Possible solutions
- Piëzo Buzzer / Speaker
- Sound generator with piëzo speaker
- Speaker directly controlled by microcontroller's PWM signal

Solution	Pro's	Con's
Buzzer	Cheap Loud	Fixed tone Only on /off
Generator and speaker	Many tone possibilities	Space consuming
Software & Speaker	Many tones and intervals possible	More Programming

Figure 4.18 Possible solutions

4.2.3.7. Optional data logging

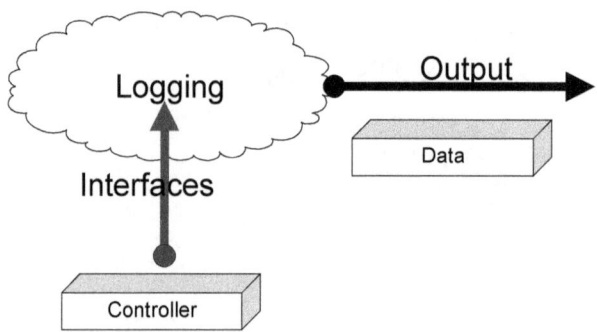

Figure 4.19 Optional data logging

Possible solutions:
- Internal storage microcontroller possible, serial readout.
- Use USB stick for storage
- Use memory card for storage (SD card)
- Wireless connection to store data on smart phone

Solution	Pro's	Con's
Intern memory	Cheap	Limited storage
USB	Large storage possible	More hardware More space needed
SD Card	Compact Large storage possible	More programming
Bluetooth & Smartphone	OEM New possibilities	Expensive More programming Always in need of smart phone

Figure 4.20

4.2.4. Concluding the Design stage

Now all partial designs have been solved, a final decision has to be made on what solutions we will use for the CREATE stage.
Since we have several options to choose from, a morphologic table comes in handy:

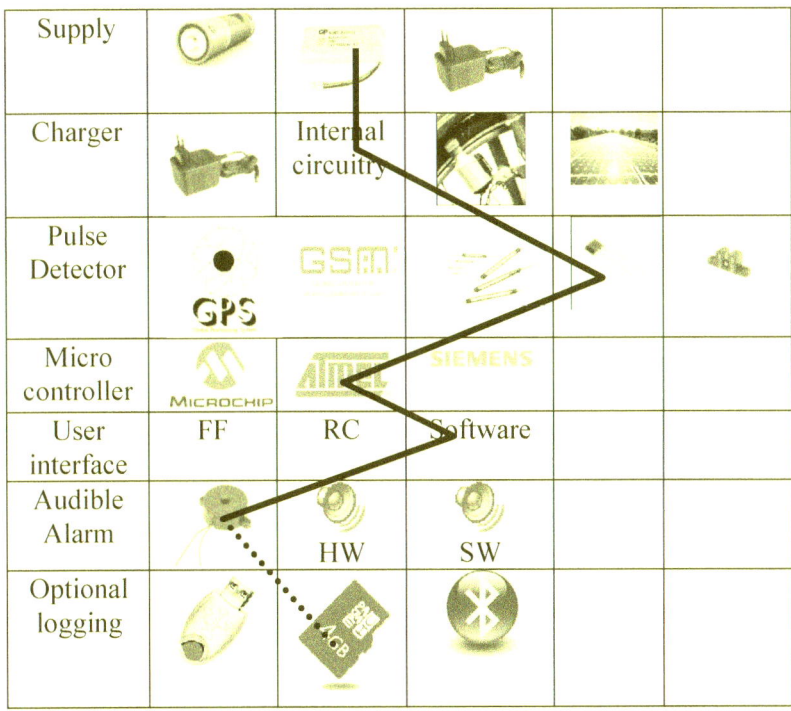

Figure 4.21 Morphologic table

4.3. Create Stage

After we made a choice about what solutions to use for the partial designs, we can now start designing these solutions during the CREATE stage.

- Power Supply
- Optional loader
- Signal transducer / converter
- Microcontroller
- User interface
- Alarm sounder
- Optional data logging

It is not necessary to design the list above in that order. It is up to the designer to decide what to do first. In fact, some partial designs are depending on others in order to function and they will have an impact on that choice.
For example, a car designer could make a choice about the number of wheels he wants to use in his design before designing the chassis.
We made a choice about what solutions to design but it is the designer's final choice to decide about the details.

4.3.1. Context diagram

Because it is essential that the designer has a complete idea of the total concept, we will start the CREATE stage with the context diagram that has been made during the SPECIFY stage.

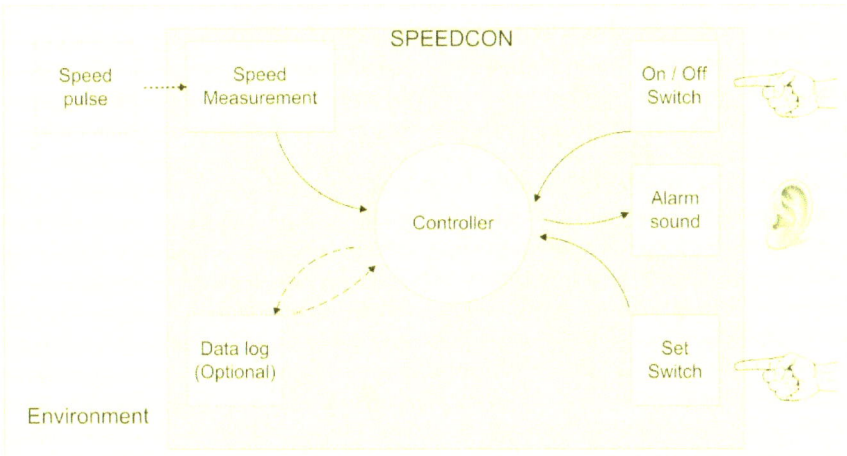

Figure 4.22 Context diagram

For a good solid design of embedded systems, it is essential that the designer knows the difference between hardware and software in regard to his task at hand. He has to make a decision about what partial designs are done by hardware, software or a combination of both. It is the combination of hardware and software that makes an embedded design powerful and interesting. For this example, the designer chose to design the hardware first and to design adapted firmware later. It is therefore important to know the limitation of both hardware and software in relation to the chosen components.

4.3.2. Creating the partial designs

4.3.2.1. Internal Power supply

During the DESIGN stage, a choice was made to use a rechargeable battery pack as power source.

The output voltage and battery capacity have not been determined at this time. In regards to the battery capacity, it is important that the user can use the device uninterrupted for at least 8 hours. He even prefers to be able to use it for 24 hours.

The chosen microcontroller will operate on about 3V, so it's logical to choose a battery that has a similar output voltage. Battery capacity of the available batteries starts at 15mAh and goes beyond the range of 9Ah.

4.3.2.2. Optional loader

The designer could construct the device in a way that enable the user to remove the power pack with the purpose to charge it externally but that is not what he had in mind.

The designer thought of a way to charge the power pack without the need to remove it.

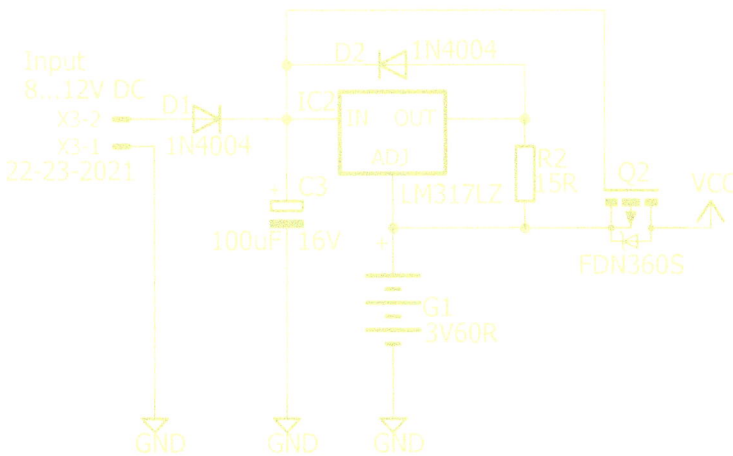

Figure 4.23 Schematic of the charger

The internal power pack can be charged by applying a power source (external adapter OEM 8..12V DC) to connector P1. Regulator IC1 is connected in a way that it operates as current source. The charging current will be equal to 1,25V/R1. F1 functions as a blockage and it will prevent the device from powering on whenever the charger is plugged in. This is not the most effective way to charge the battery but it is a rather quick and dirty solution that does not cost a lot of extra money.

4.3.2.3. Pulse detection / converter

A Hall sensor will be used to detect the spinning of the wheel so that it can be converted to speed using the software in the microcontroller. This also means that a sensor has to be attached near one of the wheels of the bike and that we mount a small magnet somewhere on the wheel. The combination of a Hall sensor and magnet will make it possible to detect the number of turn per time interval that a wheel makes during cycling.

The choice of what hall sensor to use is not limited to one. Many types and shapes are available. For this design, the designer has chosen to use a sensor type SS445P or SS461C, made by Honeywell. Of course, you can replace it by one of many alternatives but make sure you compare a few sensor specifications before you finalize your decision. The sensor should be capable of operating at the supplied voltage and the type of output is important as well. Some sensors have an open collector output combined with an internal pull-up resistor while others only have the transistor and no internal resistor. There are also versions available that have a logic gate output. The sensor that was chosen in this example has an open collector output with an internal pull-up resistor:

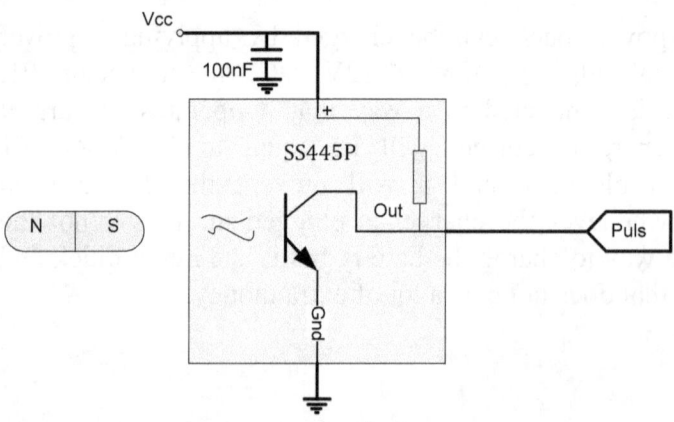

Figure 4.24

4.3.2.4. Alarm sounder

During the DESIGN stage, a choice has been made to create the needed sounds by using a buzzer. We will therefore include a piëzo buzzer that is directly connected to an output of the microcontroller.
Project Unlimited produces a buzzer AI-1223-TWT-2R that is widely available and we will include this in our project.
(Many alternatives are available and if you decide to change this component, do not forget to compare specifications first.)

4.3.2.5. User interface

To operate this device, a single button will suffice. This button can be used to set the speed that needs monitoring and the same button can be used to switch the device on and off. (Press and hold)
Electric bouncing of the switch will be compensated for in the firmware. Because the microcontroller has interval pull-up resistors for its inputs, the switch can be directly connected between the input and ground.

4.3.2.6. Microcontroller

The designer has broad experience with Atmel controllers and that is why he choose to use a microcontroller from the AVR family. Many AVT controllers can do the job and we only need minimum capacity to realize this rather simple design. The microcontroller should at least have an SPI interface so that we can use an SD card interface without too much hassle and a PWM output to drive the speaker could also safe some programming time. Not having a PWM output is not a deal breaker because we could compensate for that using firmware.
For this design, the designer chose the Atmel ATtiny2313 controller. This controller has enough flash memory to load the firmware and it has several PWM channels available. The operation voltage of this controller is 2,7 ....5,5 Volts.

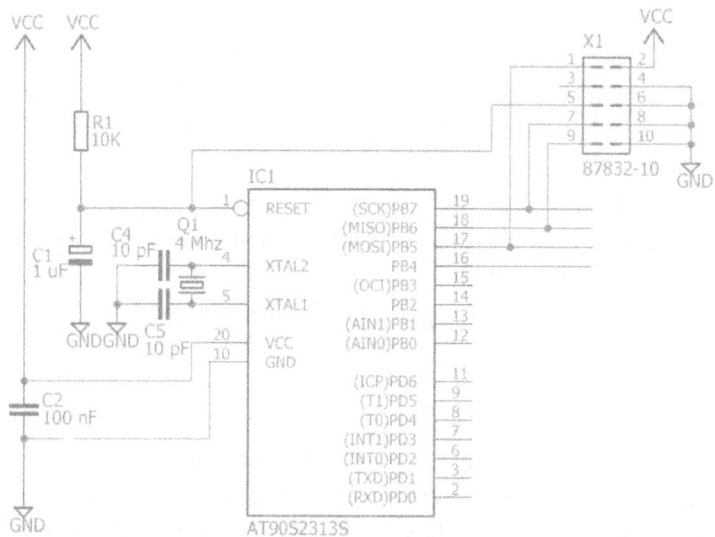

Figure 4.25 Microcontroller setup

We could design the schematics to work on a 5V power supply because the whole world of electronics is well known with this (TTL) voltage. Using 5 Volts, we would have to modify the SD interface to prevent it from breaking because the SD cards works on a maximum of 3 Volts.

It would be a better option to design the whole schematics for a 3V voltage. Doing this will not only save some components but it will also save space because we will be able to user smaller power packs.

4.3.2.7. Optional Data logging

To realize a useable data logging function, it will not be possible to make use of the microcontroller's internal memory. This can be done for short rides but it turns out to be useless for longer rides due to the large amount of generated data.

As an option in the hardware, the designer included an SD-card interface. The microcontroller will be capable of writing data directly to the card. The data on the card can then be accessed on a computer.

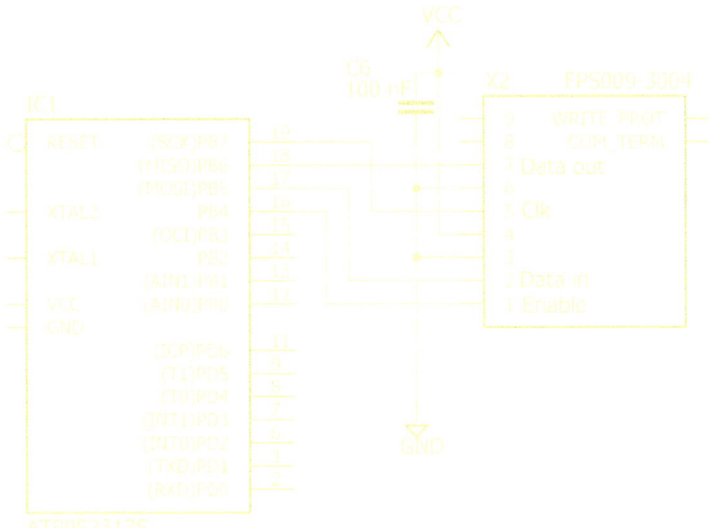

Figure 4.26 SD-card interface

However, keep in mind that the data logging was optional and it was not a mandatory requirement. So it is up to the designer if he wants to include in the data logging. This could give him extra credits, depending on his ambitions towards the customer. He could decide to include this option in the PDB design but to leave out the components during assembly. That way, he will be able to use the logging option at any given time by adjusting the assembly list.

4.3.3. Hardware

Because all partial designs have been completed, all components can be put together in one schematic:

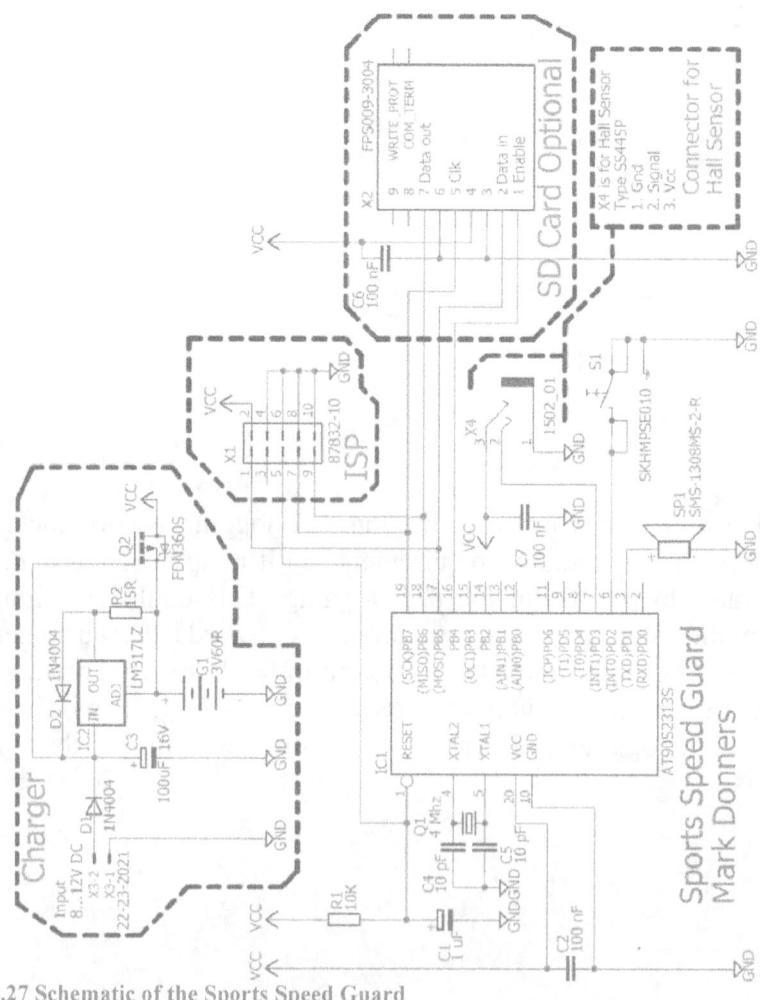

Figure 4.27 Schematic of the Sports Speed Guard

By using the right software, the designer can create a printed circuit board. (PCB). However, since this is just an example and we do not have a purpose for a real device, the PCB design steps are not actually done at this time. You can create a PCB using your own software if you wish; the schematic should be completely functional. (I did test it and a previous version of this schematic has been published by Elektor under the name 'speedcontrol' 081127-11)

4.3.4. Firmware

Often firmware is created 'on the fly' without the use of an engineered framework. This will lead from fragmented code to a complete chaos in programming. This is especially true, when you are asked to make a few adjustments to the firmware at a later time. That's why documentation is important. The designer should always make some kind of software framework that he will use as a guide during programming. There are several tools to do so. (Use Cases, flow diagrams etc.)

4.3.4.1. Use Case

The dynamic behavior of the system can be documented using *Use Case diagrams* and *Use Case descriptions*.
For this design, the designer made both.

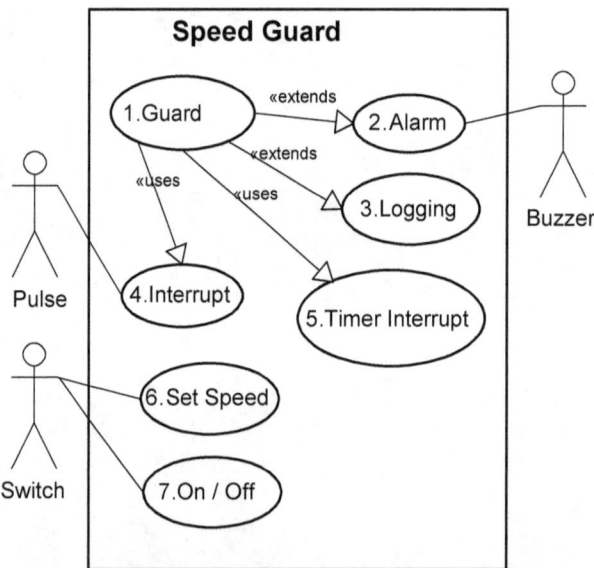

Figure 4.28 Use Case diagram

4.3.4.2. Use Case Descriptions

The use cases that were described in the *Use Case Diagram* can be described by using *Use Case descriptions*. The numbers that are used in each Use Case name related to the *Use Case diagram*.

Use Case description	
Name	1.Guard
Summary	The actual measured speed is compared to a set value. If there is a defined difference between these two, a state of alarm is activated.
Actors	None
Related to	2.Alarm 3.Logging 4.Interrupt 5. Timer Interrupt
Assumptions	System is not in sleeping mode
Description	1. Every time that the timer of "5.Timer Interrupt" causes an interrupt, a counter is increased by value '1'. The actual value of this counter is related to the actual speed. 2. Whenever a pulse is offered as a result of the "4.interrupt", A counter will be stopped or reset. 3. The instance this counter is stopped, its value represents the actual speed. 4. The actual speed is compared to the set value and the difference is calculated. 5. Depending on the amplitude of this difference, an alarm is generated by Use Case "2.Alarm". 6. The actual speed is passed on to Use Case "3.Logging".
Exceptions	-
Results	If the difference between set speed and actual speed is larger than defined, the appropriate alarm will sound.

Use Case description	
Name	2.Alarm
Summary	If addressed by Use Case "1.Guard", an alarm will be generated.
Actors	Buzzer
Related to	1.Guard
Assumptions	System is not in sleeping mode
Description	There are 3 states for this alarm: 0. Alarm off 1. Alarm: 3 short tones of 500 uSec 2. Alarm: Constant tone

Exceptions	None
Results	Incoming request is processed to change to an alarm state.

Use Case description	
Name	3.Logging
Summary	Actual speed is stored onto SD card.
Actors	None
Related to	1.Guard
Assumptions	System is not in sleeping mode
Description	As soon as the system is activated, the actual speed is stored with a 2 second interval.
Exceptions	If there is no card present, nothing will be stored.
Results	Speed data is stored onto SD card.

Use Case description	
Name	4.Interrupt
Summary	Whenever a pulse is received from the sensor, an interrupt will occur.
Actors	Pulse
Related to	1.Guard
Assumptions	System is not in sleeping mode
Description	1. If a pulse is received,(rising edge) a counter will be reset or stopped. 2. If a flag was set in Use Case "6 Set Speed", the actual speed will be set as 'set value'.
Exceptions	
Results	The "Speed set value" will be set or the actual speed will be saved to a variable.

Use Case description	
Name	5. Timer Interrupt
Summary	The timer generates an interrupt every 100uSec.
Actors	None
Related to	1. Guard
Assumptions	System is not in sleeping mode An SD card is present
Description	1. Every 100 uSec, an interval is generated because an internal timer has increased to a defined value. 2. During this interrupt, a counter is increased. 3. If this counter reaches 32000, it will stop.
Exceptions	None
Results	The counter represents the actual speed.

Use Case description	
Name	6. Set Speed
Summary	The actual speed is set as new "set speed" value.
Actors	None
Related to	1. Guard
Assumptions	System is not in sleeping mode SD card is present
Description	1. If the user pressed the switch, a bit is flagged (1 time only). As a result of this flag, Use Case "4. Interrupt" will act different.
Exceptions	None
Results	Flag is set

Use Case description	
Name	7. On / off
Summary	System can be powered on or off.
Actors	Switch
Related to	1. Guard
Assumptions	
Description	1. If the system is off (standby mode), it can be powered on by pressing the switch. 2. If the system is on (operating mode), it can be powered off by pressing and holding the switch for more than 5 seconds.
Exceptions	
Results	System toggles power mode.

4.3.4.3. Bubble diagram

Once the use cases have been completed, it would be a perfect time for the designer to make an inventory of what the microcontroller should be able to do once everything is finished. Let me clarify. We already made a choice on what microcontroller to use but we did not decide on how we will do the internal hardware and coding.
What timers will we use and in what mode will we use them? Does the microcontroller have a watchdog and are we going to use it?
Does it have power saving modes that might come in handy?
Do we need to use all internal timers or can we complete our design by only using one timer?

To answer these questions, the use of a *bubble diagram* can be useful.

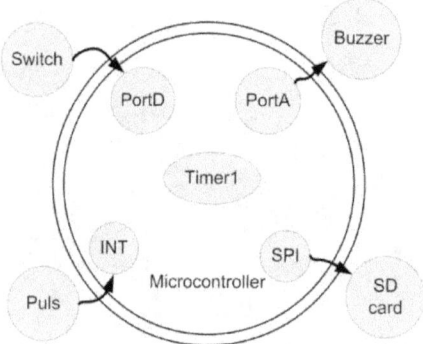

Figure 4.29 Bubble diagram

(Please take a look at the specific chapter if you cannot remember how to use the diagrams). As you will quickly discover, the use of a bubble diagram for a simple design as this one is disputable. Like with all tools, it is up to the designer to decide if he wants to use it.

4.3.4.4. Flow charts

To visualize the flow of the software, the designer can make use of *flowcharts* (flow diagrams). (The use of *Nassie Scheidermann diagrams* is another option to achieve this.)

It is possible to use the flowchart as a starting point to build state diagrams or to write code directly but often the diagrams are only used to explain the flow of the software to get a better impression of what the software is suppose to do.

Remember that converting a flow diagram to code or converting a state diagram to code will have different code results. The results will be the same but the way the coding is done will differ for sure.

As with most decisions, it is up to the designer to choice what tools to use. I would expect that the complexity of the system will have an influence on this choice.

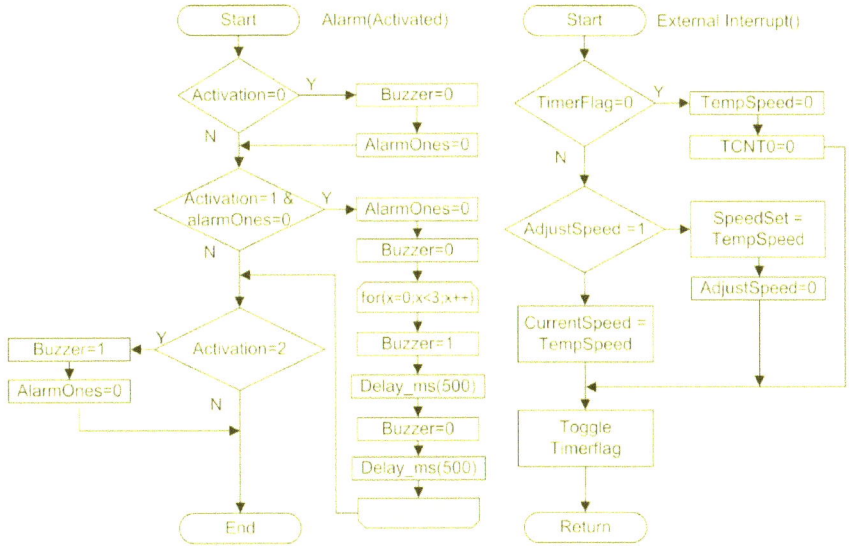

Figure 4.30 Flow chart part 1

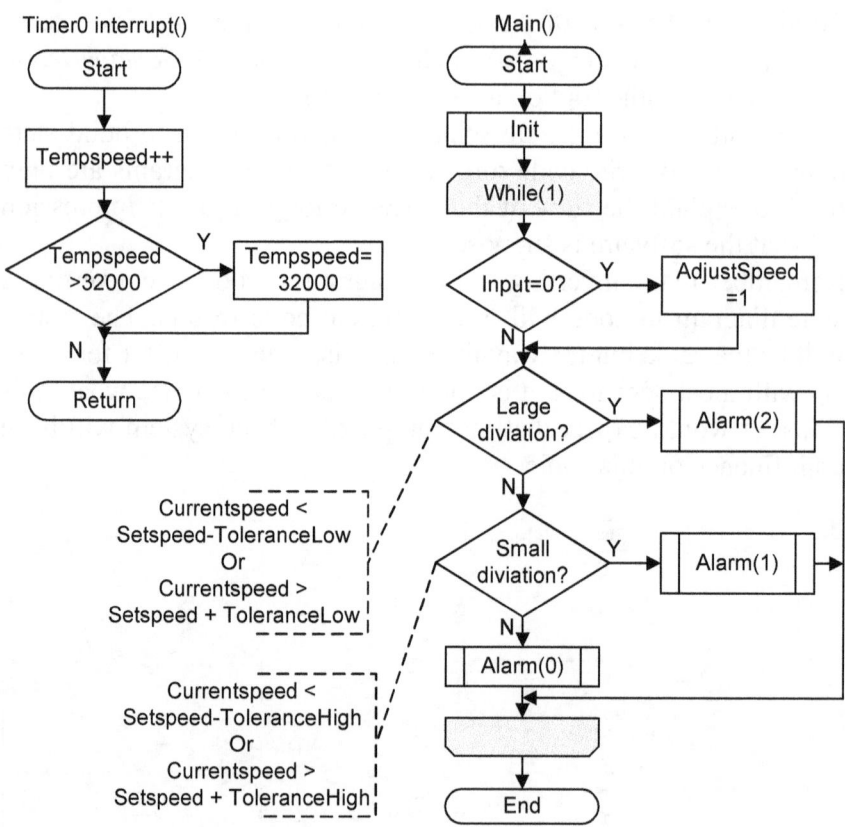

Figure 4.31 Flow chart part II

4.3.4.5. Code

The firmware as coded by the designer is listed below. Try to find the relation between flowcharts and code. (Not everything is coded because this case is an example.)

Global variables	Comments
#include <tiny2313.h>	Library
#include <delay.h>	Library
int TachoPuls=0;	
int SpeedSet=0;	Default set speed = 0
int CurrentSpeed=0;	
int TempSpeed=0;	
int SpeedTolerance=0;	
int Tolerance=0;	
bit AlarmFlag=0;	
bit AdjustSpeed=1;	flag; 1= again
bit AlarmOnes=0;	
bit TimerFlag=0;	
void Alarm(char melding);	
#define ToleranceLow	SpeedSet/8
#define ToleranceHigh	SpeedSet/16
#define Buzzer	PORTD.1
#define Input	PIND.2

Timer 1 output compare A interrupt service routine
// there was a compare match interrupt after 100 uSec
TempSpeed++; // Increase speed counter value by 1
if (TempSpeed>32000)TempSpeed=32000; // until max 32000

Power options
The functionality of using the power modes in combination with the button to toggle power modes has not been written in this example.

Logging
For this example, the data logging functionality has not been coded. The hardware is available so technically it is possible to realize it.

Initialize	Comments
PORTA=0x00; DDRA=0x00;	All input All Tristate
PORTB=0x00; DDRB=0xB0;	Bit 4,5,7=Output Rest is input tristate
PORTD=0x00; DDRD=0x02;	Bit 1 = Output Rest is input tristate
TCCR1A=0x00; TCCR1B=0x09; TCNT1H=0x00; TCNT1L=0x50; ICR1H=0x00; ICR1L=0x00; OCR1AH=0x01; OCR1AL=0x90; OCR1BH=0x00; OCR1BL=0x00;	Timer 1 initialization Clock 4 Mhz Mode:CTC top=OCR1A Noise Canceler: Off Compare A Match Interrupt: On Compare value 0x190
GIMSK=0x40; MCUCR=0x03; EIFR=0x40;	INT0: Rising Edge
TIMSK=0x40	Timer(s)/Counter(s) Interrupt initialization
ACSR=0x80;	
#asm("sei")	Enable interrupts

```
Main()
Alarm(1); // Signal a series of short beeps after activation
while (1)
  {
  /* If the switch is pressed, the flag for setting a new actual speed should
     be set. If the speed differs to much from set point, a constant tone
     should be generated
     If speed differs is small, a few short beeps should be generated
  */
    if (Input==0) AdjustSpeed=1;
    if (CurrentSpeed <  (SpeedSet-ToleranceLow) )Alarm(2);
       else
    if (CurrentSpeed <  (SpeedSet-ToleranceHigh) )Alarm(1);
       else
    Alarm(0);
  }
```

External Interrupt 0 service routine
/* By using a time TimerFlag toggle after each interrupt, one time the timer is reset and the other time the actual speed is determined. */
if (TimerFlag==0)
       {TempSpeed=0;
       TCNT0=0; // Reset timer value and current speed
       }
  else
      { if (AdjustSpeed==1)
        { SpeedSet=TempSpeed;
         AdjustSpeed=0;
        }
       else ;// currentspeed is timer value
        CurrentSpeed=TempSpeed
      }
TimerFlag=!TimerFlag; // toggle timerflag

Alarm(char melding)
Char x=0;
if(melding==0)
    {
    Buzzer=0;
    AlarmOnes=0;
    }
    if(melding==1 && AlarmOnes==0)
      {AlarmOnes=1; // flag to prevent repeat
      for(x=0;x<3;x++)
       { Buzzer=1;
        delay_ms(500);
        Buzzer=0;
        delay_ms(500);
       }
      }
    if(melding==2)
     { Buzzer=1;
      AlarmOnes=0;
     }

4.4. Validate Stage

The hardware has been built, the firmware has been written and some first testing has been completed. This means that the time has come to validate the system. Is the design compatible with all the previously defined specifications and requirements? Assuming that the customer wants to get what he paid for, it is essential to validate the system and to document this validation. Do not forget to look at any changed that have been agreed upon during the project. You can use the table of requirements and specifications to check every item. Once finished, you can show this document to the customer. Be aware! Validating is more than simply checking every item on a list. Some items need some extended testing to really be sure they are compliant.

4.4.1.1. Specifications and requirements, again.

It is essential to freshen up memory on the specifications and requirements that both designer and customer agreed upon:

Specifications and requirements	Wish / Demand	Check
General		
Operational temperature 0 °C- 40 °C	D	☐
Powered by battery	D	☐
Rechargeable	W	☐
Internal charger	W	☐
Time of operation fully charged > 8 Hours	D	☐
Time of operation fully charged > 24 Hours	W	☐
Mechanical		
Water resistant / Splash proof	D	☐
Attachable to steering wheel	D	☐
Future options		
Data recording	W	☐
CE compatible design	W	☐
Functional		
Button to set speed 'set point'	D	☐
On / off switch	D	☐
2 different alarms when speed varies	D	☐
Double function for button to eliminate on / off switch	W	☐
Logistics		
Lead time prototype < 5 months after project start	D	☐

Figure 4.32 Specifications and requirements

4.4.1.2. Test procedures

A few tests have to be completed in order to complete the validation of the system. The designer has to write a protocol to test the battery in order to see how long a single charge will last.

This could be done by performing a duration test. Will the charge last for as long as required or even longer and if so, how long? Or the designer could do some calculations to see if this specific requirement is met.

Also, the overall functionality of the device has to be tested. This should also be done by using a protocol written by the designer.

If the data logging functionality was included it would have to be tested also. (Not in this version of the firmware). Next, two examples of a test protocol are shown.

Test protocol	
Device:	Sports Speed Guard
Name of test	Functional test
Setup:	Signal generator with TTL output
Procedure:	
Remove the cable from the sensor input connected to the main circuit board. Connect the TTL signal from the signal generator to the test pin on the speed input.(make sure to also connect the ground) Start the signal generator and send a TTL square wave signal with a frequency that represents a speed 20Km per hour. Assuming that a bicycle will travel about 1 meter per wheel cycle, this would be close to 5...8 Hz. Next, press to set the speed you want monitored. The system should now have stored the current speed to be monitored. Next, lower the frequency of the signal generated by the signal generator with about 1 Hz. A few short tones should be generated to indicate that the current speed is different from the set speed. Change the frequency back to the original setting and repeat the same sequence to see if the behavior is the same as before. Next, lower the frequency even more until you hear a long constant tone. Change the frequency back to normal to see that the sound stops. Don't forget to put back the jumper Jp1 when you are finished.	

Test protocol	
Device:	Sports Speed Guard
Name of test	Battery endurance test
Setup:	Signal generator with TTL output and sweep functionality.

Procedure:
Remove jumper Jp1 from the main circuit board. Connect the TTL signal from the signal generator to the test pin on the speed input.(make sure to also connect the ground)
Start the signal generator and send a TTL square wave signal with the following settings: Sweep 4-10Hz at a speed of about 10 seconds.
Wait until the actual frequency on the generator is at approximately 8 Hz.
Next, press to set the speed you want monitored. The system should now have stored the current speed to be monitored.
Write down the current time of day to determine the time you start testing.
Check if the system is still working after 8 hours. Leave the system in this testing mode and check every 2 hours to see if it is still functional until it stops working. When it stops working, write down the actual time of day and calculate the number of hours the system was able to work on one battery charge. Don't forget to put back the jumper Jp1 when you are finished.

4.5. Finishing the case

The system is only ready if the VALIDATION stage was successful. This also means that the process of designing has been completed and aside from some administration, the project can be finished. Congratulations, you made it!

You will do yourself and all your (future) team members a huge favor by making sure all documentation is complete and properly achieved. Remember, someone might need it in the future! There is nothing more frustrating than a design that only exists in the (forgotten) memory of a single designer.

If you do not know where to start documenting, try using the design documentation that you have created during the 4 stage process. It is really a good point to start from.

5
Case Study 2: Old Riddles die hard

'A writer is someone who can make a riddle out of an answer.'
Karl Kraus

In this case we will have a go at solving an old riddle. We will create an electronic puzzle to solve one of the oldest riddles.

How can the farmer get all across the river without any loss?

5.1. Specify stage

5.1.1. Preparations

The customer who likes to collect ancient and modern toys has asked to transform an old riddle into a modern electronic puzzle.
Since you do not have much to start with, it would be a good idea to do some orientation about who you are going to meet. Find out about his background. Did he study electronics, mechanics or something else? Also, make sure you have your information ready to share. What is the average lead-time, production time, availability, knowledge within the team, and so on?

5.1.2. The interview

The customer's idea becomes clearer during the interview. It is all about an old riddle that involved as farmer who has to transport a goat, a cabbage and a wolf to the other side of a river with a small boat. The boat is only strong enough to carry the farmer and one other object/animal. The thing is that both the goat and the wolf are hungry and everything is okay as long as the farmer is supervising. However, the second he goes away, the wolf might eat the goat or the goat might eat the cabbage. So the challenge is to find a way to get the farmer with all of his things to the other side of the river in one piece.

The riddle can be solved by doing some thinking and it doesn't take a modern computer system to do so. But, according to the customer, there are people who do not find the solution that easy and if the riddle is modernized, he believes to have a market for it.

He expects to get a prototype out of this project. It should not be a modern game computer but it should be something simple, elegant and easy to operate. Therefore, a display should not be included and the prototype should have no more than buttons and lights. The reason for this is that the customer wants to integrate the electronics into a wooden puzzle. The prototype can be powered by batteries. Other than the things mentioned above, the customer has no opinion about how it should be realized.

5.1.3. Brainstorm

Although the problem at hand is rather compact, finding the right solutions can be challenging. The overall design should be appealing enough to get the users' attention. When they see it, they should want to start playing with it.

This is enough reason to sit down with a bunch of people to exchange ideas. Together with the customer and 2 electronic engineers several ideas have been put together. The result was a mind map illustration as shown below.

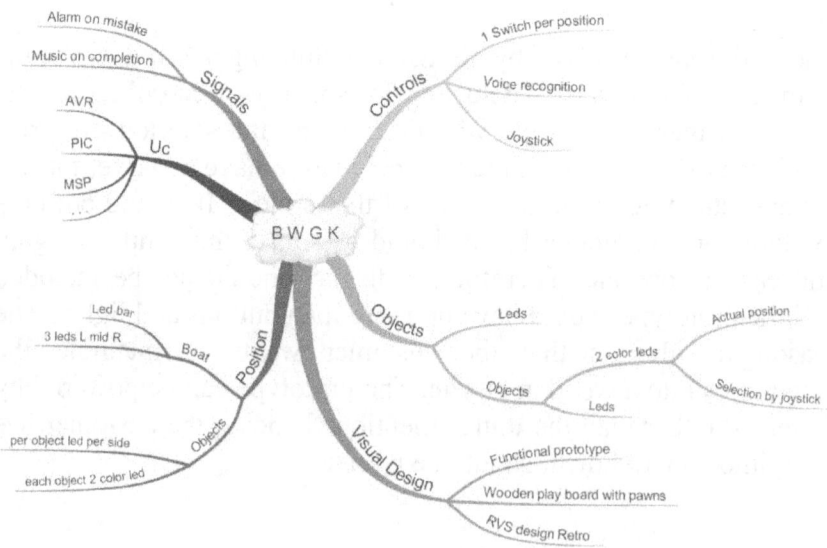

Figure 5.1 Mind map created with iMindMap www.ThinkBuzan.com.

This is the right time to give this product a name that can be referred to during the whole development process: BWGK.

5.1.4. Product sheet

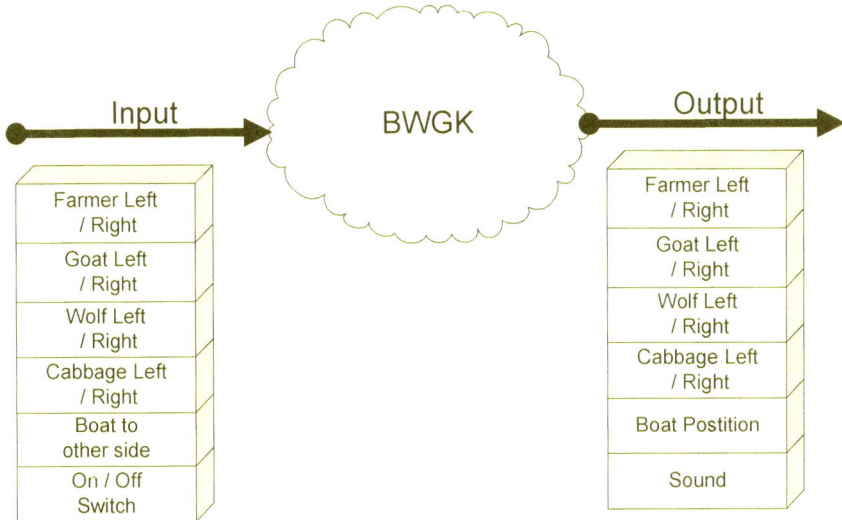

Figure 5.2 Black-box design of system BGWK

Function: Puzzle, toy. Tool to solve riddle.

The illustration above shows a first impression of the overall design. Although, the information in the illustration might suggest something different, the mind map that was created earlier has no influence on this illustration. So far, no choice in components has been made. The illustration nearly shows that we have some objects as inputs and some as outputs.

5.1.5. Converting to specifications

Specifications and requirements	Wish / Demand	Check
General		
Powered by battery	D	☐
Mechanical		
Prototype in functional case	D	☐
Future		
Design of case	W	☐
CE compliant	W	☐
Functional		
User interface for objects and boat	D	☐
Power switch	D	☐
Alarm sound on mistakes in solving puzzle	D	☐
Melody sound when puzzle is solved	W	☐

Figure 5.3 Requirements and specifications.

After putting all the requirements and specifications in writing, it is a good time to show them to the customer and ask for his approval. A good way of doing this is to include this list in the quotation that he is going to sign.

5.1.6. Context diagram

The SPECIFY stage will be concluded with an overall context diagram.
The diagram clearly shows what is part of the system (green) and what is part of the environment. (Blue).

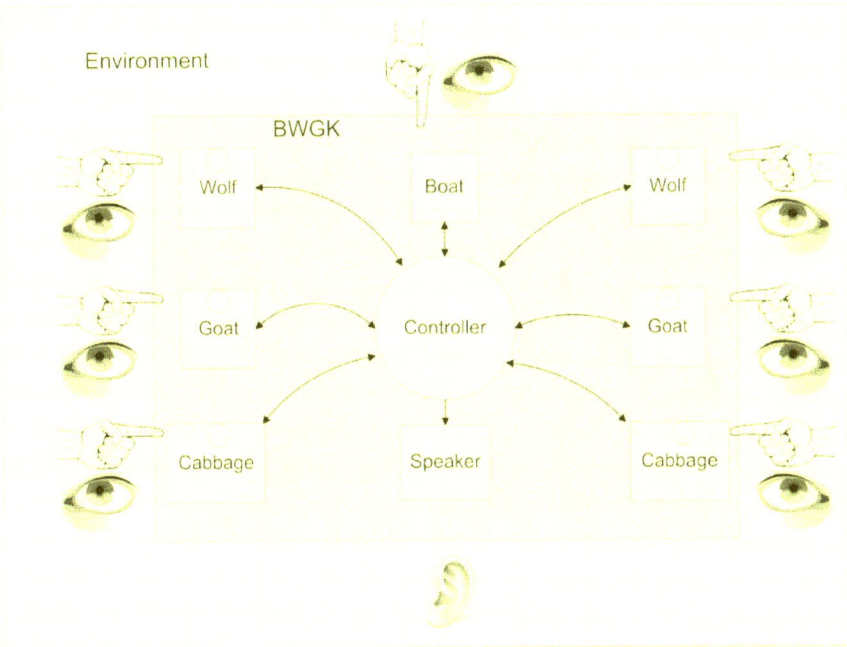

Figure 5.4 Context Diagram of BWGK system

5.2. Design Stage
5.2.1. Defining the system

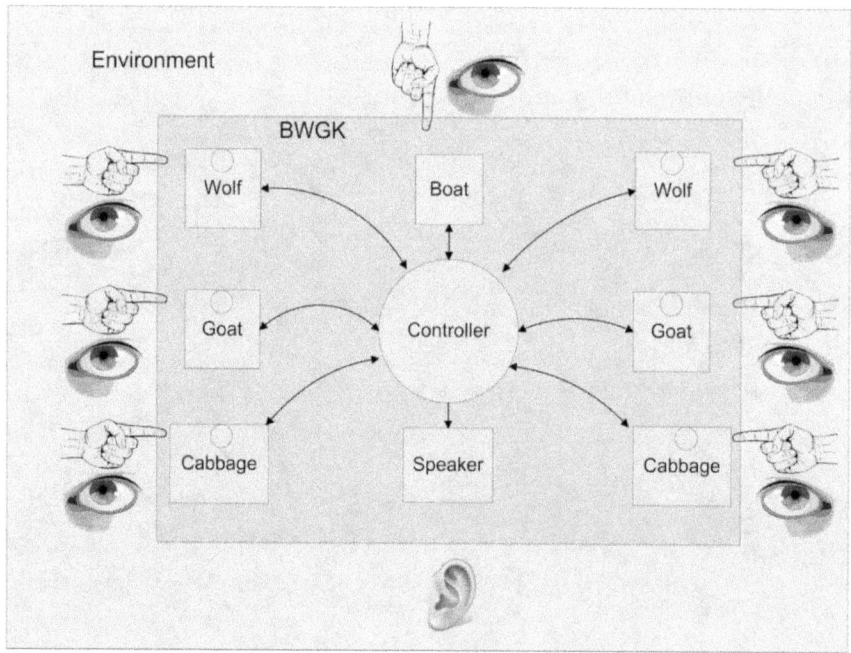

Figure 5.5 Context diagram

The context diagram shows that we need to design a system called BWGK. The user interface can be clearly distinguish within the environment and consists of several single inputs, outputs and some sort of speaker.

Let's take another look at the black-box cloud representation of the system to refresh our memory:

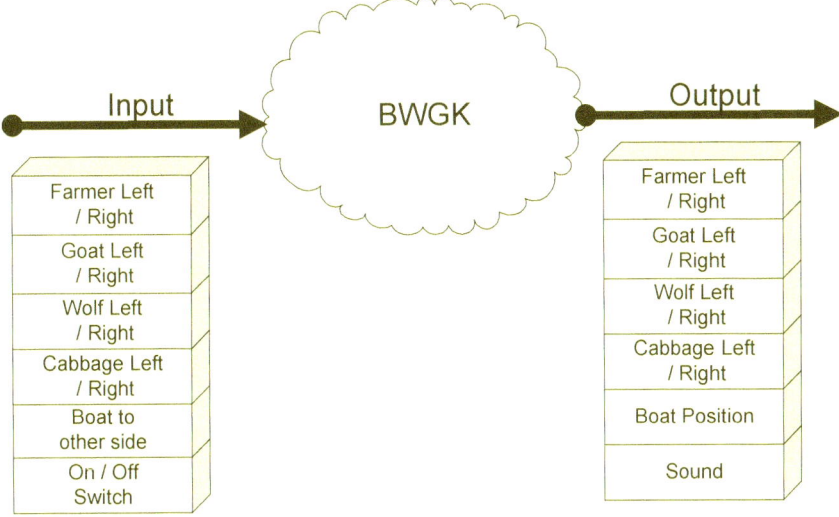

Figure 5.6

5.2.2. Splitting up into partial designs

The black-box of this system can be split up into several partial designs. For this design the following partial designs are described:

- Supply
- Micro controller
- User interface (Anti bounce)
- Indicators
- Sounder

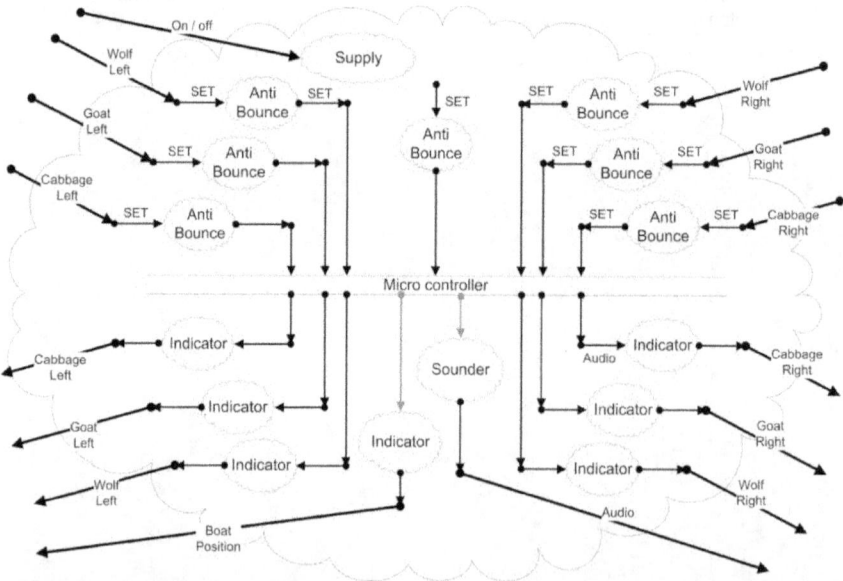

Figure 5.7 Inside the black box

The design cloud from the previous illustration shows that the system contains many similar partial designs. All indicators are alike except the one in the middle that is used for the boat position. Indicating the boat position can be done by a series of LEDS or the use of a display.
The anti bounce partial designs are also equal. Anti bounce can be achieved in many ways and is often done by a combination of hardware and software. But solving the anti bounce, is not a concern at this time.

5.2.3. Defining the partial designs.

In this part of the DESIGN stage, the partial designs will be described further. The level of details is limited because no choice of components has been made final. Sometimes, not adding details is difficult. Take this case for example; the designer has not decided about what technologies he is going to use but he is already talking about using led bars to indicate to position of the boat?
Do not let a design method stop you from expressing great idea's but remember, it is always nice to see what else is out there.

5.2.3.1. Supply

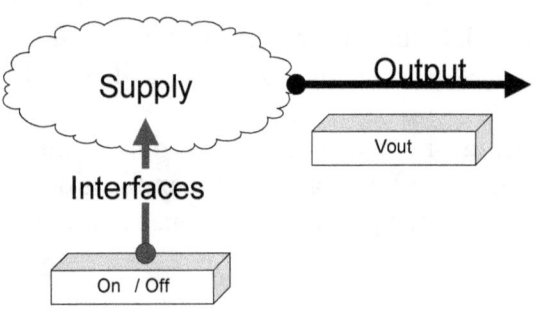

Figure 5.8

One of the specifications and requirements was to use a battery as power supply. So, we only need to think about what battery to use and how to convert the battery voltage to a working voltage for the microcontroller. (if applicable)

Possible Solutions:

- Penlight batteries 1,5V per battery, combine several ones.
- 9V battery met regulator

Solution	Pro's	Con's
Penlights	Widely available Large capacity	Space consuming if more than 2
9V Battery	Widely available Reasonably in size	Less capacity Needs voltage reduction circuitry to be used with microcontroller.

Figure 5.9

5.2.3.2. Microcontroller

Again, the choice of micro controller is in the hands of the designer. All has to do with previous experience and availability in development tools.

As mentioned before, it is often wise to use a microcontroller that is related to ones that have already been used in previous designs.

It is a way of reducing the expenses because all tools for development are already within the company.

Below a few possible solutions are shown.

Solution	Pro's	Con's
PIC	Known to the designer	Limited tools available within the company
ATMEL AVR	Known to the designer	All tools available within the company
MSP430	Cheap	All tools available within the company

Figure 5.10

5.2.3.3. User interface

For this design, all user controls are alike with the exception of the on / off switch. These controls are limited to several single switches.

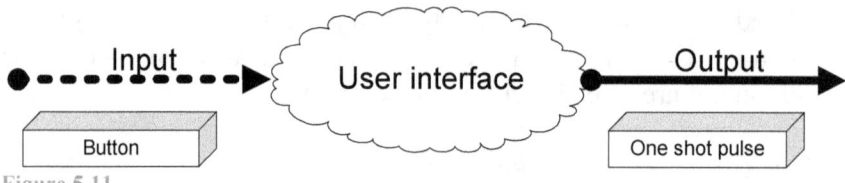

Figure 5.11

Possible solutions:

- Anti Bounce hardware by using one shot flip flop
- Anti Bounce hardware by using RC network combined with software coding.
- Software coding only

Solution	Pro's	Con's
One shot FF	Reliable	Extra components
HW & SW	Reliable	Extra Coding needed
Software	Reliable	Extra Coding needed

Figure 5.12

Optional function for power toggle by pressing one or more buttons for longer period of time.

Solution	Pro's	Con's
Coding	No power switch needed	More Coding needed

Figure 5.13

5.2.3.4. Sounder

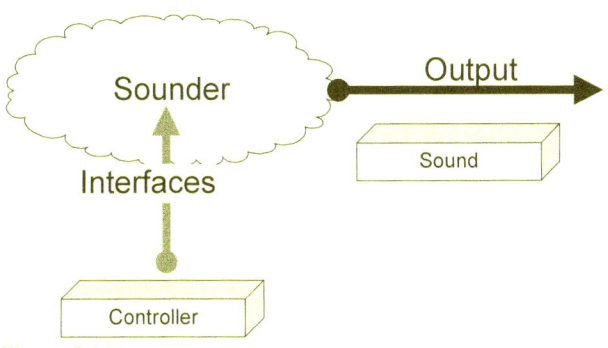

Figure 5.14

Possible solutions:
- Buzzer

- Generator with speaker or piëzo-element

- Speaker controller by microcontroller

Solution	Pro's	Con's
Buzzer	Cheap	Fixed tone only on / off
Generator & Speaker	Unlimited Tones	Space consuming
Software & Speaker or piëzo	Unlimited tones and patterns	Extra coding needed

Figure 5.15

5.2.3.5. Indicators

Like the user controls, the indicators are also alike. There are several options but the most convenient one would be the use of low current LEDS. However, to indicate the position of the boat, we could think of alternatives.

Indicating Three positions would be enough. (Two for each river side and one for the river itself.) It could be fun to indicate more positions in order to simulate the boat crossing the river.

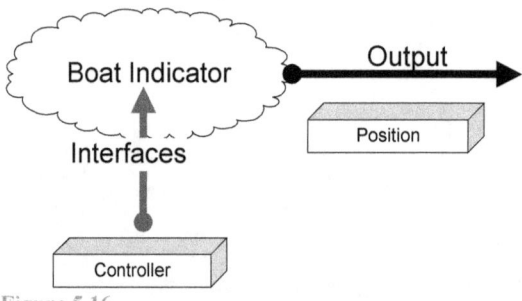

Figure 5.16

Possible solutions:
- Multiple LEDS, LED Bar

- 3 LEDS : Left side, Water, right side

- Use the horizontal elements of a 7 Segment LED display

Solution	Pro's	Con's
LEDBAR	Animation possible	Many micro controller outputs or encoder needed
3 LEDS	Low budget	No animation
7 seg. display	Low budget	7 sig. display

Figure 5.17

5.2.4. Concluding the Design stage

Now that for every partial design, one or more solutions have been listed, it is time to make a choice on which ones will be used in the CREATE stage.

Since we listed several solutions for every partial design, the use of a morphologic table is useful.

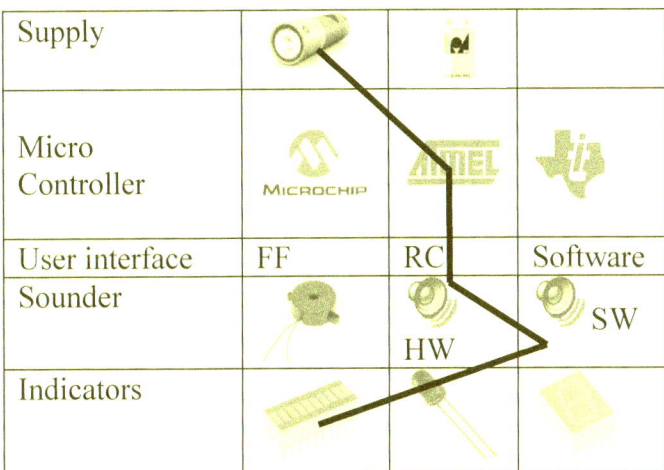

Supply			
Micro Controller	MICROCHIP	ATMEL	TI
User interface Sounder	FF	RC HW	Software SW
Indicators			

Figure 5.18 Morphologic table

The line shows the partial designs that will be created during the next stage.

5.3. Create Stage

During this Stage the chosen partial design will be created and designed on a hardware and software level.

5.3.1. Context diagram

Again, we will start by taking another look at the systems context diagram.

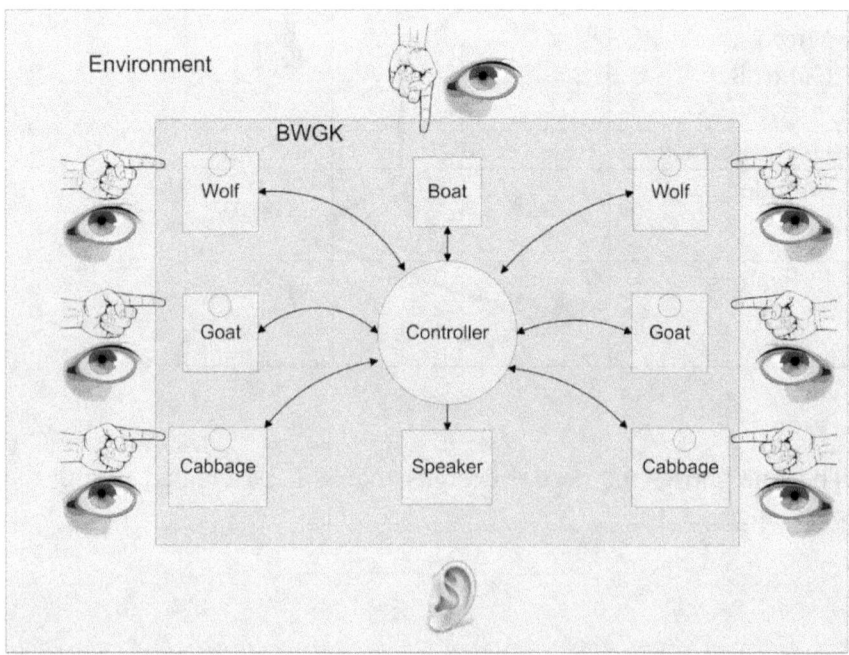

Figure 5.19 Context diagram

5.3.2. Creating the partial designs

5.3.2.1. Supply

The Atmel AVR microcontroller family that was chosen by the designer also has controller versions that can operate on 2 to 4 penlight batteries. (3V..5V). This makes a partial design of some sort of power reduction unnecessary. Power can be toggled on and off by a switch that is directly connected to the battery

5.3.2.2. Microcontroller

The microcontroller that will be used in this design is an Atmel AVR controller type ATTINY2313V that can operate at a supply voltage as low as 1,8V. It has an internal oscillator that makes the need of an external resonator unnecessary. Not using the external resonator gives us two extra inputs or output channels at our disposal. To be able to use these extra channels, some fuses have to be programmed to tell the controller that we will use the internal oscillator.

5.3.2.3. User interface

The user interface consists of several buttons that can be used to transfer object with the boat across the river if the required conditions are met.
Basically, we should use some sort of anti bounce to prevent the switches from electric bouncing when they are pressed.
During the DESIGN stage, we chose to use some form of animated LED bar to indicate the position of the boat. These LEDS will be animated showing the boat traveling across the river. After a switch is pressed, the procedure of displacing the boat starts. The boat traveling across the river will take longer than one second and during that time, all switches have no function. This is why we do not have a need for anti bounce measures. Although during the DESIGN stage we have chosen to use an RC anti bounce circuitry, we have now changed this decision into using no anti bounce

measures. This is up to the designer to decide but since this is deviating from the original plan, documenting such a decision is essential.

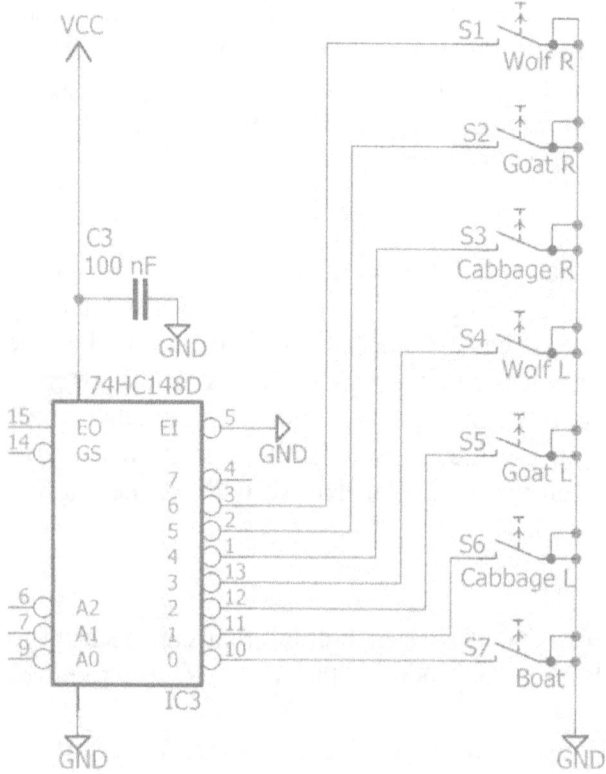

Figure 5.20 User interface Switch encoder

Since we have seven switches; six objects and one boat, we will also need seven inputs on the microcontroller. However, because we are also using several outputs, we do not have enough I/O available on the micro controller.

This is why we use an encoder to encode the inputs to a three bits number. This is done by using a 74LVC138. The LVC version of this chip can operate on the supplied voltage of 3 volts.

5.3.2.4. Sounder

To be able to play any victory tunes or defeat melodies, the use of a simple monotone buzzer will not do. This is why we use a small piëzo speaker that we will activate with a pwm signal generated by the microcontroller directly. This speaker is therefore directly connected to the microcontroller. This might not be the most efficient way of doing it but it will do for this purpose.

5.3.2.5. Indicators

Six LEDS will be used to giveaway the position of each object. These LEDS will each be connected to the microcontroller using a serial resistor to limit the LEDS' current.

To indicate the position of the boat, we will use a cluster of four LEDS. All LEDS are connected with a serial resistor that will limit the current for the connected LED.

5.3.3. Hardware

Now that all partial designs have been created, it is time to put them together in an overall design.

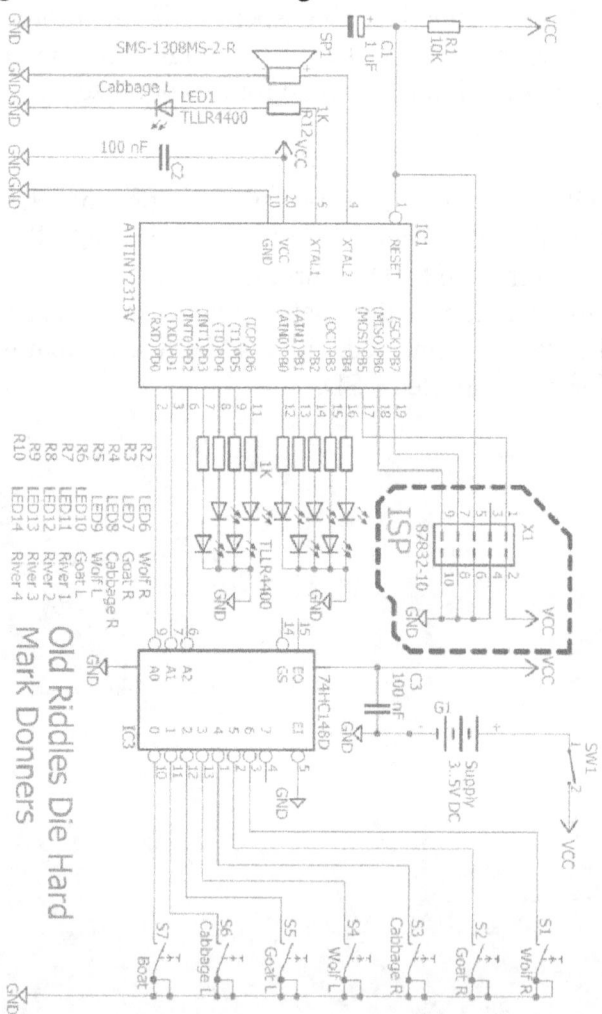

Figure 5.21. Schematic

5.3.4. Firmware

5.3.4.1. Use Case

We will describe the behavior of the system by using Use Case diagrams.

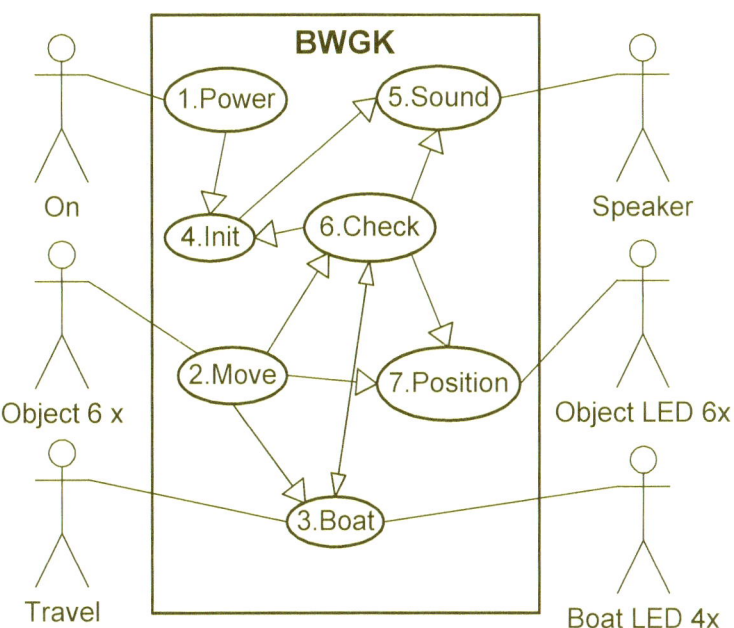

Figure 5.22 Use Case diagram

5.3.4.2. Use Case Descriptions

The Use Case Diagram can now be described by Use Case descriptions.

Use Case description	
Name	1.Power
Summary	This case takes care of changing the power state that the device is in.
Actors	On
Related to	4.Init
Assumptions	-
Description	The ON switch can be used to toggle the systems power.
Exceptions	None
Result	Device changes power state

Use Case description	
Name	7.Position
Summary	Reveals the actual position of the objects
Actors	Object LED(6x)
Related to	6.Check,2.Move
Assumptions	-
Description	Each object has a LED indicator on both sides of the river to indicate its actual position. This position can be changed by Use Case MOVE. The Use Case CHECK can also put the LED indicators in a state of alarm to indicate that the rules have been broken or the game has been solved.
Exceptions	
Result	Object is indicated in its actual position.

Use Case description	
Name	2.Move
Summary	An object can be moved from one side to the other by pressing the Object Switch.
Actors	Object(6x)
Related to	6.check,7.Position,3.Boat
Assumptions	To be able to move an object, it has to be on the same side of the river as the boat. Only the switch on the side that the object is on will have result.
Description	When an object switch of an object is pressed, the object will move to the other side of the river. For this movement, the case POSITION will be used to indicate the movement of the boat. The case CHECK will be used to verify that the journey is compliant with the rules of the game.
Exceptions	-
Result	If the journey is compliant with the rules of the game, the object will end up on the other side of the river.

Use Case description	
Name	3.Boat
Summary	This case will enable the boat to travel to the other side of the river when the travel button is pressed.
Actors	Travel, Boat LED(4x)
Related to	2.Move, 6.Check
Assumptions	-
Description	1.When the user pressed the TRAVEL button, the boat will travel to the other side of the river. This is indicated by a bitshift on the LEDBAR which indicated the position of the boat using Use Case MOVE 2. If indicated by the Use Case MOVE, action number 1 can be started. 3. The Use Case CHECK can decide to enable an alarm stage for all LEDS whenever the rules of the game are broken or when the game has been solved. 4. Use Case CHECK will check if the rules are broken during traveling.
Exceptions	-
Result	Boat on other side of the river. Or game has reset.

Use Case description	
Name	4.Init
Summary	After booting and after a restart that initiated by solving the game, the initial parameters are loaded that will reset the system.
Actors	-
Related to	1.Power,5.sound,2.Check
Assumptions	-
Description	1.The Use Case POWER indicated that the game has to be started. This will call the Use Case INIT to restore all initial values and reset the system. The Use Case SOUND wil play the starting sound from the speaker 2.If asked by Use Case CHECK, the Use Case INIT will be initiated to restore all initial values and reset the system.
Exceptions	-
Result	The game has been reset to initial values

Use Case description	
Name	5.Sound
Summary	If certain conditions are met, a sound will be produced.
Actors	Speaker
Related to	4.Init, 6.Check
Assumptions	-
Description	1. After the game has booted and loaded as indicated by the Use Case INIT, the speaker will produce a sound. 2. If one of the rules of the game are broken, the Use Case CHECK will sound an alarm and the game will be reset by the Use Case INIT. 3. The Use Case CHECK will indicate if the game is solved and a melody will be produced by Use Case SOUND. After that, the game will be reset by the Use Case INIT.
Exceptions	-
Result	Sound will be produced by speaker

Use Case description	
Name	6.check
Summary	
Actors	-
Related to	4.Init,5.Sound,2.Move,3.Boat,7.Position
Assumptions	-
Description	If requested by a related case, the CHECK Use Case will validate the actual position of the objects and the boat to see if the rules of the game are broken. If the rules are broken, the Use Case SOUND wil send an alarm to the speaker and the Use Case INIT will reset the game to the initial values. The Use Case CHECK can also command the Use Case BOAT to move the boat to the other side of the river. The Use Case check can also order the Use Case POSITION to Flash the LEDS to give signal that the rules have been broken.
Exceptions	
Result	An object or the boat has moved to a new position or the game has been reset.

5.3.4.3. Code

For the purpose of this example, the complete code is not included. Instead, code fragmentation of the user cases is included to show that Use Case diagrams and descriptions can be used perfectly for object oriented coding. (A working code for the BWGK hardware is available on request. All you need to do is send me an email to get it.) Take note that no flow-diagrams or NS-diagrams have been made. The code below is a direct translation of the Use Case descriptions.

```
Code for Use Case 1:Power
// System powers up from start and the initial values will be loaded
#include <90s2313.h>
#include <delay.h>
#define CabbageL PORTA.0
#define CabbageR PORTB.2
#define GoatL PORTB.0
#define GoatR PORTB.3
#define WolfL PORTB.1
#define WolfR PORTB.4
#define Speaker PORTA.1
#define River1 PORTD.3
#define River2 PORTD.4
#define River3 PORTD.4
#define River4 PORTD.5
#define Button PORTD & 0x07
#define left 1
#define waterL 2
#define waterR 3
#define Right 4
#define off 0
#define on 1
#define FAIL 1
#define VICTORY 2
#define BOOT 3
```

Code for Use Case 2: Move, combined with Use Case 7: position
Void Move(){ // move an object to other side of river
// 0=boatL=cabbL,2=goatL,3=wolfL,4=cabbR,5=goatR,6=wolfr,7=default

```
Switch(Button){
  case 0: if (BOATPOS==left)MOVE(right);
          else MOVE(left) //Boat to other side
          break;
  case 1: if(CABBAGEPOS==left & BOATPOS==left){
            CabbageL=0; // led off
            Move(right);
            CabbageR=1; //Led on
            CABBAGEPOS=right;
          }
          break;
  case 2: if((GOATPOS==left & BOATPOS==left)){
            GoatL=0; // led off
            Move(right);
            GoatR=1; // led on
            GOATPOS=right;
          }
          break;
  case 3: if((WOLFPOS==left & BOATPOS==left)){
            WolfL=0; // led off
            Move(right);
            WolfR=1; // led on
            WOLFPOS=right;
          }
          break;
  case 4: if(CABBAGEPOS==right & BOATPOS==right){
            CabbageR=0; // led off
            Move(right);
            CabbageL=1; //Led on
            CABBAGEPOS=left;
          }
          break;
  case 5: if(GOATPOS==right & BOATPOS==right){
            GoatR=0; // led off
            Move(right);
            GoatL=1; //Led on
            GOATPOS=left;
          }
          break;
```

```
  case 6: if(WOLFPOS==right & BOATPOS==right){
          WolfR=0; // led off
          Move(right);
          WolfL=1; //Led on
          WOLFPOS=left;
          }
       break;
  case 7: break;        // default
  }
}
```

Code for Use Case 3: Boat
```
Void Move(char direction)
{
   If(direction== right){ // let's move from left to right
   River1=off;
   Delay_ms(300);
   River2=on;
   Delay_ms(300);
   River2=off;
   Delay_ms(300);
   River3=on;
   Check(); // let's see if rules of the game are broken
   Delay_ms(300);
   River3=off
   Delay_ms(300);
   River4=on;
   }
If(direction== left){ // let's move from right to left
   River4=off;
   Delay_ms(300);
   River3=on;
   Delay_ms(300);
   River3=off;
   Delay_ms(300);
   River2=on;
   Check(); // let's see if rules of the game are broken
   Delay_ms(300);
   River2=off
   Delay_ms(300);
   River1=on;
   }
}
```

Code for Use Case 4: Init
Void Init(){
// all object to start position, LEDS on left side of river will be on, rest will be off
char BOATPOS = left;
char CABBAGEPOS = left;
char GOATPOS = left;
char WOLFPOS = left;

CabbageL = on;   CabbageR = off;
WolfL = on;   WolfR = off;
GoatL = on;   GoatR = off;
River1 = on;   River2 = off;   River3 = off;   River4 = off;
}

Code for Use Case 5: Sound
Void Sound(char type){ // FAIL=1,VICTORY=2,BOOT=3
 int duration;
  Switch(type){
  Case 1: for(duration=1;duration<300;duration++){
        Delay_us(800);
        Speaker=!Speaker;
        }
        for(duration=1;duration<300;duration++){
        Delay_us(800);
        Speaker=!Speaker;
        }
        for(duration=1;duration<300;duration++){
        Delay_us(800);
        Speaker=!Speaker;
        }
        Break;
  Case 2: for(duration=1;duration<300;duration++){
        Delay_us(1200);
        Speaker=!Speaker;
        }
        for(duration=1;duration<300;duration++){
        Delay_us(1100);
        Speaker=!Speaker;
        }
        for(duration=1;duration<300;duration++){
        Delay_us(1000);
        Speaker=!Speaker;
        }
        for(duration=1;duration<300;duration++){

```
            Delay_us(900);
            Speaker=!Speaker;
            }
            for(duration=1;duration<300;duration++){
            Delay_us(800);
            Speaker=!Speaker;
            }
            for(duration=1;duration<300;duration++){
            Delay_us(700);
            Speaker=!Speaker;
            }
            for(duration=1;duration<300;duration++){
            Delay_us(600);
            Speaker=!Speaker;
            }
            for(duration=1;duration<300;duration++){
            Delay_us(500);
            Speaker=!Speaker;
            }
            Break;
    Case 3: for(duration=1;duration<600;duration++){
            Delay_us(800);
            Speaker=!Speaker;
            }
            Break;
    }
    Check(); // to see if the game was solved
}
```

```
Code for Use Case 6: Check
Void Check(){
  / let's see if the rules of the game are not broken
  Char temp;
  char error = 0;
  if(WOLFPOS==left & GOATPOS==left)error = 1;
  if(GOATPOS==left & CABBAGEPOS==left)error = 1;
  if(GOATPOS==right & CABBAGEPOS==right)error = 1;
  if(WOLFPOS ==right & GOATPOS==right)error = 1;
    if(error==1){ // ok rules are broken!
      Sound(FAIL);
      Init() // let's reboot
    }
  // ok no rules broken but is the game solved?
  If (WOLFPOS==right & GOATPOS==right &CABBAGEPOS=right){
    Sound(VICTORY);
      for (temp=0;temp<15;temp++){
      PORTD=0x00;
      Delay_ms(300);
      PORTD=0xFF;
      Delay_ms(300);
      }
    Init(); // lets restart
   }
}
```

The code of the use cases can be used to complete the firmware that is needed to control the hardware of the device. There are compilers available that can do the coding for you.

5.4. Validate stage

Validating the design to check if it lives up to the specifications and requirements is exactly what needs to be done in this stage. So the first thing to do is to check the specifications and requirements table.

Specifications and requirements	Wish / Demand	Check
General		
Powered by battery	D	●
Mechanical		
Prototype in functional case	D	●
Future		
Design of case	W	●
CE compliant	W	-
Functional		
User interface for objects and boat	D	●
Power switch	D	●
Alarm sound on mistakes in solving puzzle	D	●
Melody sound when puzzle is solved	W	●

Figure 5.23 Specifications and requirements table

Of course, some more testing can be defined for this validation. You could do a field test by lending the device to some people to play with.
Or you could do a battery test to see how long the battery will last. (The actual CE testing has not been done because we do not have an actual device to test in this imaginary example).

6
Case 3: Embedded Sports Trainer

'My trainer don't tell me nothing between rounds. I don't allow him to. I fight the fight. All I want to know is; did I win the round? It's too late for advice.'
Muhammad Ali

In this case, we will develop a device that can be used for interval training. Not all stages will be covered. In fact, covering the create stage is totally up to you. It is time to apply everything you have learned.

The device consists of a remote and multiple wireless receivers. The coach can press buttons on the remote to activate the remote receivers. On activation of a receiver, its light or buzzer will be activated to get the groups' attention. One way of using that is to make the group run towards it.

6.1. Specify Stage

6.1.1. Preparations

Besides my love for electronics and music, I also have a passion toward judo sports. As coach and athlete, you can find me in the gym on a regular base. The idea behind this device has its origin in the gym. Because the designer and the customer are the same person in this case, an interview is not the next logical step. So instead, a textual description of the device I have in mind will have to do.

6.1.2. Functionality of the system

There are many forms of training and one of them is running from point A to point B within a given amount of time. Football players use it to run from one corner of the field to another. A known variation to this run is the so called shuttle run test in which the time interval decreases in steps.

The design will use a microcontroller as a remote control to enable a maximum of 4 remote receivers. These receivers will be placed on the field by the coach. For example, he could place one on every corner of the field. The coach can activate a receiver so that its light and buzzer gives a signal to the athletes. The system can also use a pre-defined program like a shuttle run test or cooper test. (Twelve minutes of running your longest possible distance)

In the illustration below, you can see a mind map of the system. The parts of receiver and transmitter can be clearly distinguished.

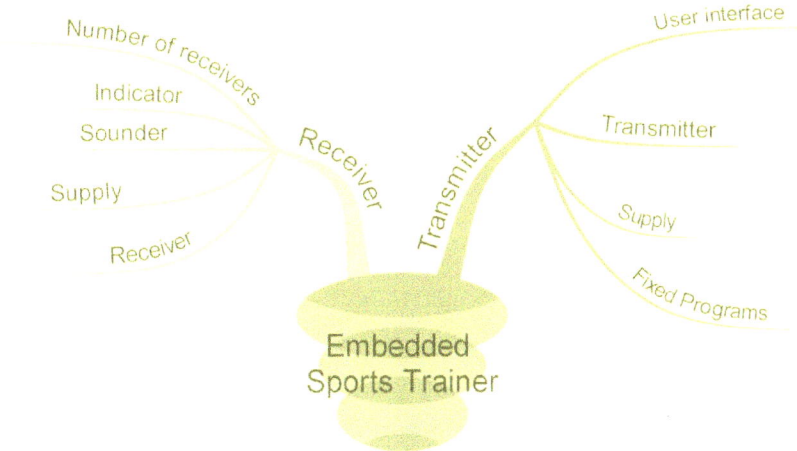

Figure 6.1 Created with iMindMap. www.ThinkBuzan.com.

6.1.3. Product sheet

The black box of the system could look like this:

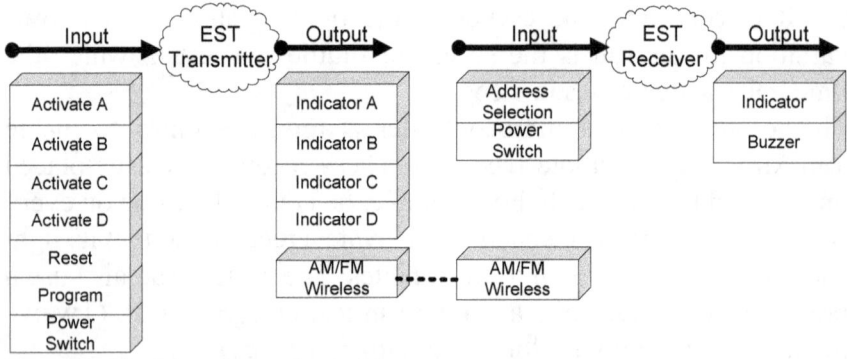

Figure 6.2 Black box design cloud of the system

6.1.4. Specifications and requirements

Even though I am serving my own table, sticking to a list of specifications and requirements keeps me on the right path and hopefully stops me from making many changes during the design process.

Specifications and requirements	Wish / Demand	Check
Transmitter		
Range more than 1 football field	D	☐
Portable	D	☐
Powered by Battery	D	☐
Receiver(s)		
Water resistant for use outdoors	D	☐
Light beacon as indicator	D	☐
Create sound on activation	W	☐
Maximum of 4 receivers	D	☐
Address of receivers easily programmed	D	☐
Powered by battery	D	☐
Power switch	D	☐
Functional		
Activate receivers selectively	D	☐
Autonomous activation of receivers by pre programming.	W	☐
2 pre installed programs (Random Run) and (Shuttle Run)	W	☐

Figure 6.3 specifications and requirements

6.1.5. Context Diagram

Of course, we will conclude the SPECIFY stage with a context diagram.

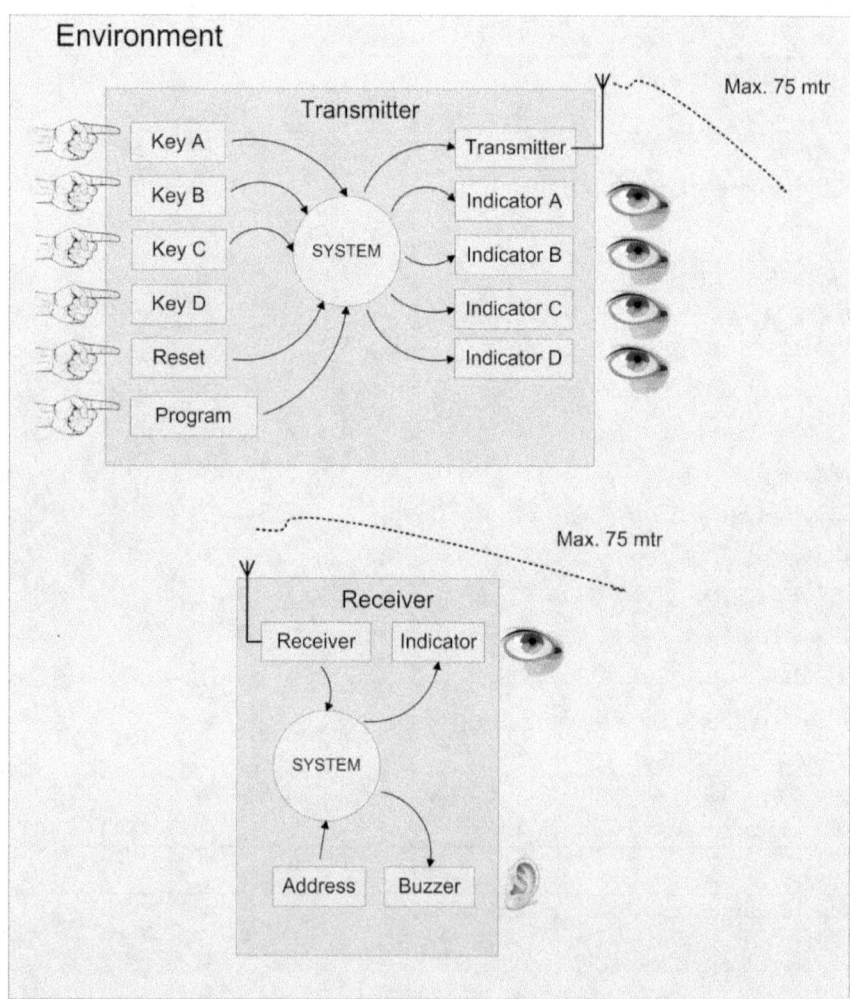

Figure 6.4 Context Diagram of the Embedded Sports Trainer EST

6.2. Design Stage

6.2.1. Defining the system

Again we will start by looking at the context diagram. This is only to refresh our memory. (In a real project there is often a longer period of time between the development stages). (You can find this diagram at the previous paragraph).

We also made a black box cloud representation of the system, here it is again:

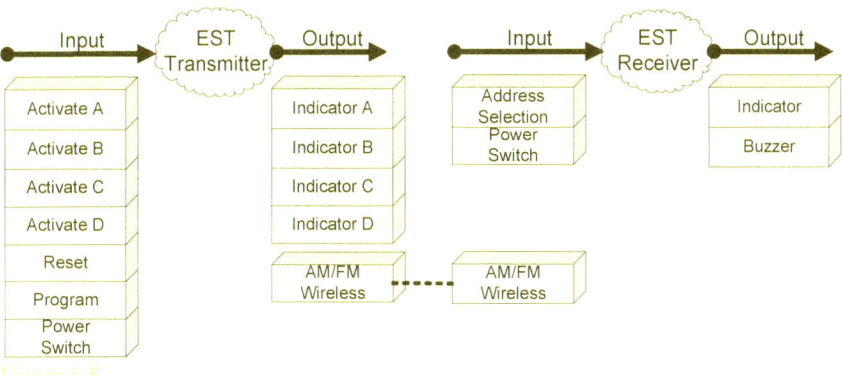

Figure 6.5

The cloud shows that we are dealing with more than one devices. We have the remote and we have a receiver. In fact, the receiver is not one device but we will build four of them.
On the receiver side, there is an input labeled "Address Selection". This is used to give every receiver a unique address. So, it all depends on the hardware and software but theoretically, the number of receivers is nearly unlimited. However, we will stick to four of them.

6.2.2. Splitting up into partial designs

The black-box of the ESD should be split up into several partial designs. I have chosen to specify the following partial designs:

Transmitter
- Supply
- Micro controller
- User interface (Anti Bounce)
- Transmitter
- Indicators

Receiver
- Supply
- Address selection
- Receiver
- Buzzer
- Indicators

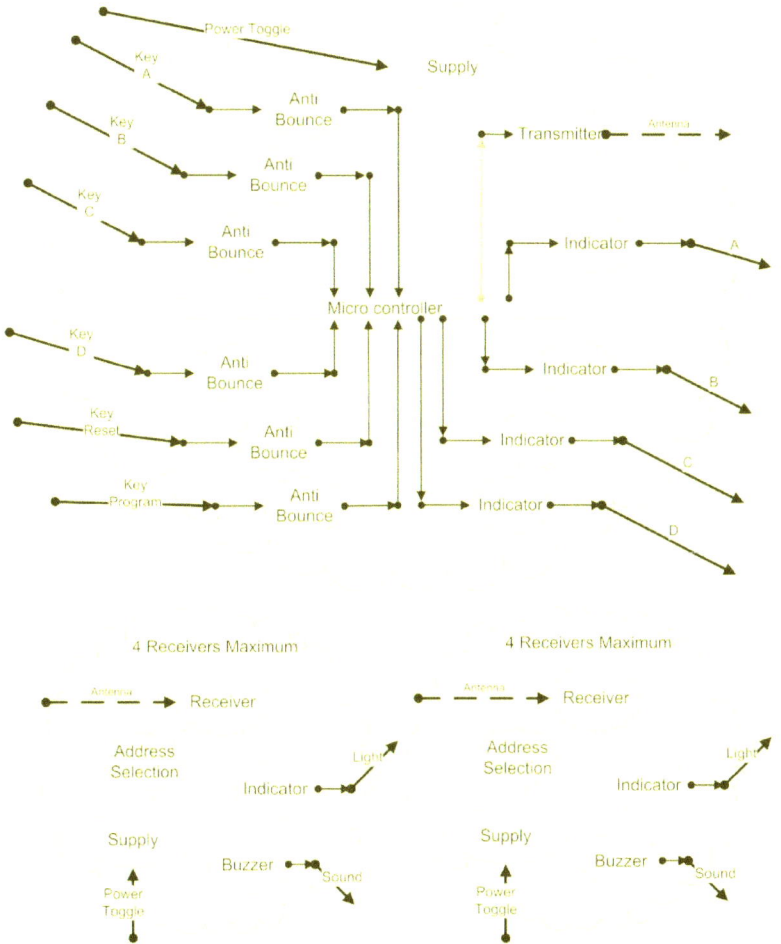

Figure 6.6

The illustration above shows again that the system consists of several devices. (One transmitter and maximum of four receivers) Unlike the other cases, this one also uses a cloud without microcontroller. This is because I did not think I would need a microcontroller to realize the receivers. (Not really a neutral perspective is it?)

6.2.3. Defining the partial designs

Now that it has been decided what partial designs will be defined, it is time to do so.

6.2.3.1. Supply

Both the transmitter and the receiver(s) must have some sort of power supply. Because we want to be able to use the device in the field, a (rechargeable) battery is the most logic option to use.

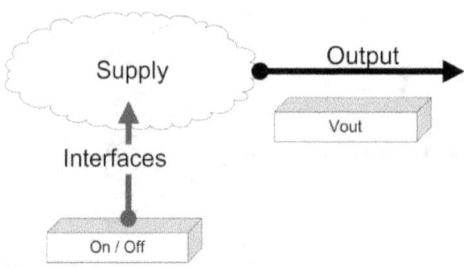
Figure 6.7

So all we need to do is to figure out what kind of batteries to use and maybe how we can convert the supply voltage to our needs. Presumably, the transmitter will have a microcontroller and a transmitter. The receiver does not necessarily need a microcontroller. The easiest way would be that all components inside the transmitter operate on the same supplied voltage. The same goes for the receiver.

Possible solutions

Solutions	Pro's	Con's
AA size batteries	Availability Capacity	Can be space consuming
9V Battery	Availability Reasonably small	Voltage conversion necessary

Figure 6.8

6.2.3.2. Transmitter / Receiver modules

There are many options available to realize wireless transmission. Some are listed below:

Solutions	Pro's	Con's
IR	Inexpensive	Limited range
Bluetooth	Open standard	Expensive implementations
FM OEM Module With chipset	Inexpensive easy to use	OEM depends on 3th party
Zigbee	Extended protocol	Difficult to implement

Figure 6.9

6.2.3.3. Microcontroller

It is reasonable to assume that only the transmitter will have a microcontroller. The choice of what microcontroller to use is up to the designer. Any microcontroller can be used as long as it operates under the right conditions and it has a few I/O. Based on experience, I have chosen to use a micro controller from Atmel.

One of the reasons why I like to stick with a microcontroller from the same family is that it will save me money. I do not need to buy another set of development tools.

A list of some other microcontrollers is shown below:

Solutions	Pro's	Con's
PIC	Known to designer	Limited tools available in house.
ATMEL AVR	Known to designer	All tools available in house
MSP430	Cheap	All tools available in house

Figure 6.10

6.2.3.4. Transmitter user interface

All user controls for controlling the individual channels are the same and is limited to a single button per channel.

Figure 6.11

Possible Solutions:
- Hardware Anti bounce with one shot Flip Flop

- Hardware Anti bounce with RC network combined with software

- Anti bounce with software only

Solutions	Pro's	Con's
One shot FF	Reliable	Extra components
HW & SW	Reliable	Extra Coding needed
Software	Reliable	Extra Coding needed

Figure 6.12

Optional control for power toggle by holding keys for a period of time:

Solutions	Pro's	Con's
Coding	No need for power switch	Extra coding needed

Figure 6.13

6.2.3.5. Buzzer

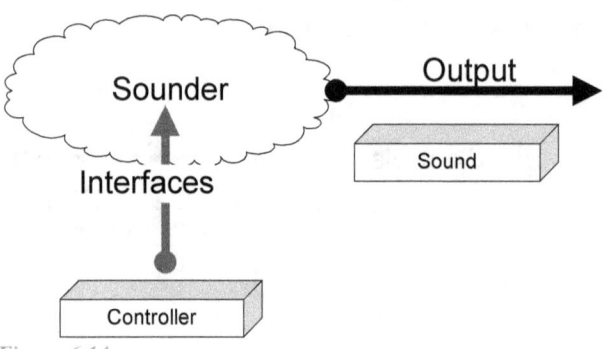
Figure 6.14

Possible solutions are listed below:

Solutions	Pro's	Con's
Siren	Inexpensive Loud	Operating voltage
Buzzer	Inexpensive	Not so loud

Figure 6.15

6.2.3.6. Indicators

We have to deal with two kinds of indicators. The transmitters will have small indicators to see what channel is activated or what program is running. LEDS would probably be the best option for those.

Each receiver will have one indicator that should be visible from a far distance.

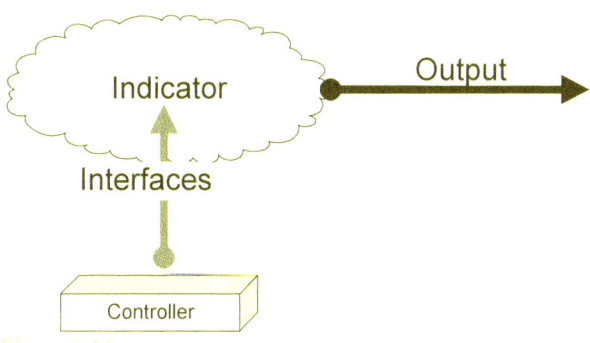

Figure 6.16

Possible solutions are listed below:

Solutions	Pro's	Con's
LED	Inexpensive Durable	Amount of light is limited
Light bulb	Cheap	Operating voltage and power
Signal light	OEM product	
LED display	Cheap	Driver needed

Figure 6.17

6.2.4. Concluding the Design Stage

Now all partial designs have been defined, the next step is to make a choice on which ones will be designed in the next stage.

A morphologic table is shown below:

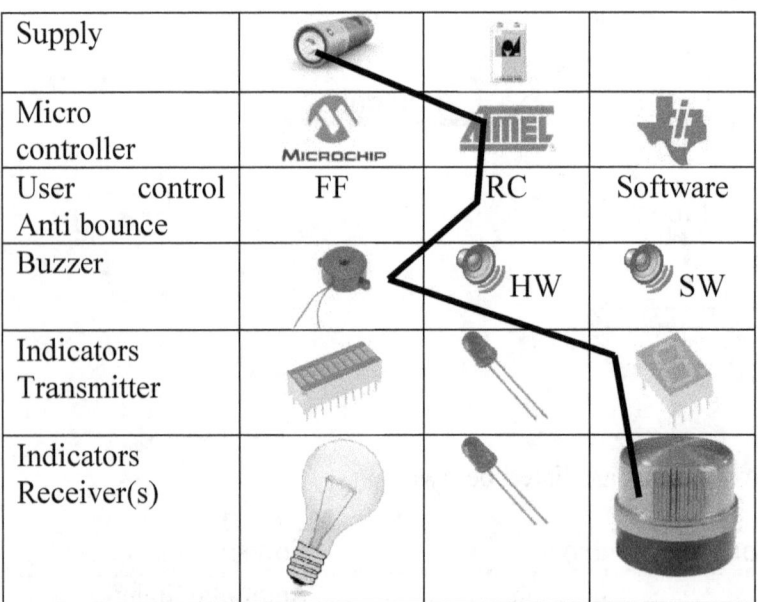

Supply			
Micro controller	MICROCHIP	ATMEL	
User control Anti bounce	FF	RC	Software
Buzzer		HW	SW
Indicators Transmitter			
Indicators Receiver(s)			

Figure 6.18 Morphologic table

The line shows the selection that has been made by the designer. However, this is where the project is put on hold. Not because I run out of paper but because I want you to apply everything you have learned by finishing the design yourself. Feel free to change the choices that are shown in the morphologic table above.

6.3. Next Stages, Create and Validate

Okay, I have taken you this far and I believe it is time for you to put your knowledge into practice. It's entirely up to you to finish the design of the embedded sports trainers.

I look forward to read your feedback and design solutions. Feel free to send me an email. You can find the email address in the first pages of the book.

Make sure you do not forget an important rule;

'Design with a KISS'.
(KEEP IT SAFE AND SIMPLE),

7
The End

I hope this book will help you in your quest to become a great designer. The book is far from complete because a good book is never finished. Also, I know that some tools in the toolbox are not following the rules it has been designed for. I tried to keep it simple, that's the truth. And please, do not let rules stop you from tweaking a tool to your needs.

'Some of the greatest inventions come from unexpected results after breaking some rules'.
Mark   Donners

Notes

Notes

Notes